TRANSPORTING COMPRESSED DIGITAL VIDEO

THE KLUWER INTERNATIONAL SERIES IN ENGINEERING AND COMPUTER SCIENCE

TRANSPORTING COMPRESSED DIGITAL VIDEO

by

Xuemin Chen
San Diego, CA

KLUWER ACADEMIC PUBLISHERS
Boston / Dordrecht / London

Distributors for North, Central and South America:
Kluwer Academic Publishers
101 Philip Drive
Assinippi Park
Norwell, Massachusetts 02061 USA
Telephone (781) 871-6600
Fax (781) 681-9045
E-Mail < kluwer@wkap.com>

Distributors for all other countries:
Kluwer Academic Publishers Group
Distribution Centre
Post Office Box 322
3300 AH Dordrecht, THE NETHERLANDS
Telephone 31 78 6392 392
Fax 31 78 6546 474
E-Mail < services@wkap.nl>

Electronic Services < http://www.wkap.nl>

Library of Congress Cataloging-in-Publication Data

A C.I.P. Catalogue record for this book is available
from the Library of Congress.

Printed on acid-free paper.

Printed in the United States of America

Contents

Preface

The purpose of *Transporting Compressed Digital Video* is to introduce fundamental principles and important technologies used in design and analysis of video transport systems for many video applications in digital networks.

In the past two decades, progress in digital video processing, transmission, and storage technologies, such as video compression, digital modulation, and digital storage disk, has proceeded at an astounding pace. Digital video compression is a field in which fundamental technologies were motivated and driven by practical applications so that they often lead to many useful advances. Especially, the digital video-compression standards, developed by the Moving Pictures Expert Group (MPEG) of the International Organization for Standardization (ISO) and the International Electrotechnical Commission (IEC), have enabled many successful digital-video applications. These applications range from digital-video disk (DVD) and multimedia CDs on a desktop computer, interactive digital cable television, to digital satellite networks. MPEG has become the most recognized standard for digital video compression. MPEG video is now an integral part of most digital video transmission and storage systems. Nowadays, video compression technologies are being used in almost all modern digital video systems and networks. Not only is video compression equipment being implemented to increase the bandwidth efficiency of communication systems, but video compression also provides innovative solutions to many related video-networking problems.

The subject of *Transporting Compressed Digital Video* includes several important topics, in particular video buffering, packet scheduling, multiplxing and synchronization. Reader will find that the primary emphasis of the book is on basic principles and practical implementation architectures. In fact, much of the material covered is summarized by examples of real developments and almost all of the techniques introduced here are directly applicable to practical applications.

This book takes a structured approach to video transporting technology, starting with the overview of video transporting and video compression techniques and working gradually towards important issues of video transporting systems. Many applications are described throughout the book. These applications include the video transporting techniques used in the broadband communication systems such as the digital broadcasting system for cable television and the direct satellite broadcasting system for digital television; transporting schemes for digital head-end multiplexing and re-multiplexing system, video transcoding system, and also the rate-control schemes for the video transmission over networks, and much more. The book is compiled carefully to bring engineers, video coding specialists, and students up to date in many important modern video-transporting technologies. I hope that both engineers and college students can benefit from the information in this, for the most part, self-contained text on video transport systems engineering.

The chapters are organized as follows:

Every course has its first lecture a sneak preview and overview of the technologies to be presented. Chapter 1 plays such a role. Chapter 1 provides an overview of video transporting systems that is intended to introduce the transport-packet multiplexing functionality and important issues related to video-transport for digital networks.

Chapter 2 provides the reader with a basic understanding of the principles and techniques of image and video compression. Various compression schemes, either already in use or yet to be designed, are summarized for transforming signals such as image and video into a compressed digital representation for efficient transmission or storage. This description of video-coding framework provides most of the tools needed by the reader to understand the theory and techniques of transporting compressed video.

Chapter 3 introduces concepts of compressed video buffers. The conditions that prevent the video encoder and decoder buffer overflow or underflow are derived for the channel that can transmit a variable bit rate video. Also, strategies for buffer management are developed from these derived conditions. Examples are given to illustrate how these buffer management ideas can be applied in a compression system that controls both the encoded and transmitted bit rates.

Chapter 4 discusses the techniques of system clock recovery for video synchronization. Two video-synchronization techniques are reviewed. One technique measures the buffer fullness at the receiving terminal to control the decoder clock. The other technique requires the insertion of time stamps into the stream at the encoder. The focus in this chapter is on the technique of video synchronization at decoder through time stamping. MPEG-2 Transport Systems is used as an example to illustrate the key function blocks of the video synchronization technique. A detailed analysis on digital phase-locked-loop (D-PLL) is also provided in this chapter.

In Chapter 5, methods for generating the Presentation Time Stamps (PTS) and Decoding Time Stamps (DTS) in the video encoder are discussed. In particular, the time-stamping schemes for MPEG-2 video are introduced as examples. It is the presence of these timestamps and the correct use of the timestamps that provide the facility to synchronize properly the operation of the video decoding.

In Chapter 6, conditions for preventing decoder buffer under-/over-flows are investigated by using the encoder timing, decoding time stamps and dynamics of encoded-picture size. Some principles on video rate-buffer management of video encoders are studied. Both MPEG-2 and MPEG-4 video buffer verifiers are also introduced in this chapter.

In Chapter 7, the discussions are focused on analyzing buffer, timing recovery and synchronization for video transcoder. The buffering implications of the video transcoder within the transmission path are analyzed. The buffer conditions of both the encoder and transcoder are derived for preventing the decoder buffer from underflowing or overflowing. The techniques of regenerating timestamps in transcoder are also discussed.

Chapter 8 devotes to topics of transport packet scheduling and multiplexing. Again, MPEG-2 transport stream target decoder is introduced as a model for studying timing of the scheduler. Design requirements and techniques for statistical multiplexing are also discussed in this chapter.

Two applications of video transport multiplexer are introduced in Chapter 9 to illustrate many design and implementation issues. One application is an MPEG-2 transport stream multiplexer in encoder and other is an MPEG-2 transport re-multiplexer.

Certain materials provided in Chapters 1, 6 and 8 are modified from or related to ATSC (A/53), ISO (MPEG-1, -2 and -4) and ITU (H.261, H.262, and H.263) standards. These standard organizations are the copyright holders of the original materials.

This book has arisen from various lectures and presentations on video compression and transmission technologies. It is intended to be an applications-oriented text in order to provide the background necessary for the design and implementation of video transport systems for digital networks. It can be used as a textbook or reference for senior undergraduate-level or graduate-level courses on video compression and communication.

Although this text is intended to cover most of the important and applicable video transporting techniques, it is still far from complete. In fact, we are still far from a fundamental understanding of many new video compression techniques, nor has coding power been fully exploited in the modern video compression systems.

I wish to acknowledge everyone who helped in the preparation of this book. In particular, the reviewers have made detailed comments on parts of the book which guided me in the final choice of content. I would also like to thank Professors I. S. Reed and T. K. Truong for their continuing support and encouragement. I also gratefully acknowledge Mr. Robert Eifrig, Dr. Ajay Luthra, Dr. Fan Ling, Dr. Weiping Li, Dr. Vincent Liu, Dr. Sam Narasimhan, Dr. Krit Panusopone, Dr. Ganesh Rajan, and Dr. Limin Wang for their contributions in many joint patents, papers and reports which are reflected in this book. I would also like to thank Mr. Jiang Fu for reading of parts of the manuscript and for thoughtful comments. It was important to be able to use many published results in the text. I would like to thank the people who made possible of these important contributions.

Support for the completion of the manuscript has been provided by the Kluwer Academic Publishers, and to all I am truly grateful. In particular I truly appreciate the attentiveness that Mr. Alex Greene and Ms. Melissa Sullivan have given to the preparation of the manuscript.

The author dedicates this work in memory of professor Fang-Yun Chen, one of the greatest Chinese contemporary scientists, for the inspiration he provided to all students, and to the practitioners of communication theory and systems.

Finally, I would like to show great appreciation to my wife, daughter, and parents for their constant help, support and encouragement.

1 Digital Video Transport System

1.1 Introduction

In the past two decades, progress in digital video processing, transmission, and storage technologies, such as video compression, digital modulation, and digital storage disk, has proceeded at an astounding pace. Especially, the video-coding standards, developed by the Moving Pictures Expert Group (MPEG) of the International Organization for Standardization (ISO) and the International Electrotechnical Commission (IEC), have enabled many successful digital-video applications. These applications range from digital-video disk (DVD) and multimedia CDs on a desktop computer, interactive digital cable television, to digital satellite networks. MPEG has become the most recognized standard for digital video compression and delivery. MPEG video is now an integral part of all digital video transmission and storage systems [1-1] [1-2].

Without question, digital video compression is the most important enabling technology for modern video communication. Enormous research and development efforts in video compression have led to the important advances in digital video transmission systems. Digital video technology brings many great advantages to the broadcasting, telecommunications and networking, and computer industries. Comparing with analog video, the use of compressed digital video provides lower costs in video distribution, increases the quality and security of video, and allows for interactivity. Some advantages of digital video compression are illustrated in the following examples. Firstly, digital compression enables a cable television system operator to carry several (e.g. four to six) television programs on one traditional cable television channel that used to carry only one service. Secondly, with compressed digital video, several (e.g. four or more) television programs can be carried on one satellite transponder that used to distribute a single channel. This results substantial saving on renting transponders. Thirdly, analog video collects noise, e.g. snow and ghosts, as it travels over the air and through the cable to homes. With error-correction technology, digital video on the other hand arrives exactly as it was sent, sharp, clear, and undistorted.

Just as there are techniques on how best to compress digital video, there are also efficient methods to manage, transmit, store and retrieve compressed digital video. Among these techniques, no one need be reminded of the importance, not only of the speed of transmission, but also of the accuracy and flexibility of the video transport process. The general term, video transport, involves packetizing, multiplexing, synchronizing and extracting of video signals. The digital revolution has provided a new way of transporting video. It also has the potential to solve many other problems associated with timely, cost-effective delivery of high-quality video and audio.

This book addresses the issues of transporting compressed digital video in modern communication systems and networks. In the text, the subject of digital video transporting is described in a practical manner for both engineers and students to understand the underlying concepts of video transporting, the design of digital video transmission systems, and the quantitative behavior of such systems. A minimal amount of mathematics is introduced to describe the many, sometimes mathematical, aspects of digital video compression and transporting. The concepts of digital video transporting are in many cases sufficiently straightforward to avoid theoretical description.

This chapter provides a description of the functionality and format of digital video transport systems. While tutorial in nature, the chapter overviews transporting issues related to the digital video delivery.

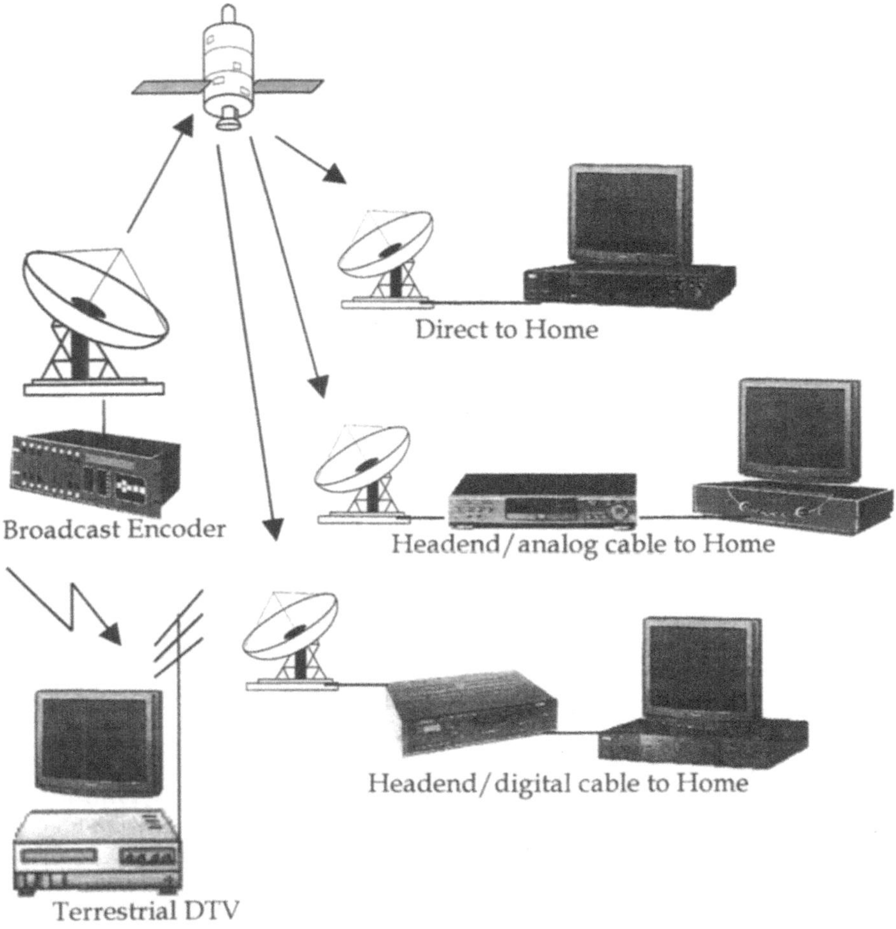

Figure 1.1. Four types of DTV networks

1.2 Functions of Video Transport Systems

To illustrate functional requirements of video transport systems, examples of digital video transmission systems are given in this section. Fig. 1.1 shows four types of satellite and cable digital television networks. These are (1) the terrestrial digital television (DTV) service, (2) a hybrid of digital satellite service and digital cable service, (3) a hybrid of digital satellite service and analog cable service, and (4) the direct satellite DTV service (DSS).

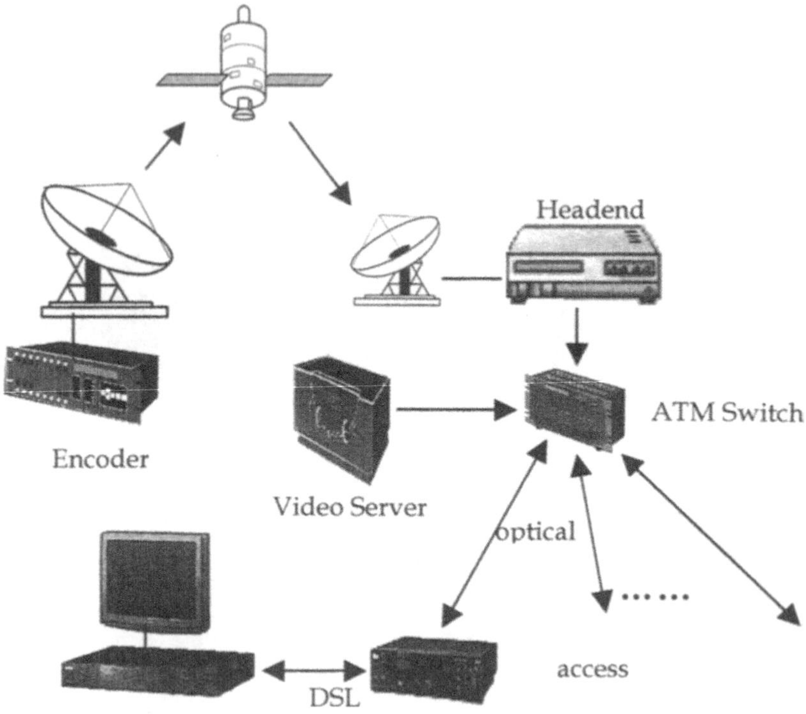

Figure 1.2. Digital video delivery over ATM network.

Multiple digital video and audio are compressed in the corresponding encoders and coded bit streams along with their ***timing information*** are ***packetized*** and ***encrypted*** and ***multiplexed*** into a sequence of packets as a single string of bits. The channel encoder then transforms the string of bits to a form suitable for transmission over a channel through some form of modulation. For example, the QPSK modulation is used in the satellite transmission and the VSB modulation is employed in terrestrial DTV transmission while the QAM modulation is applied in cable transmission. The modulated signal is then transmitted over the communication channels, e.g. through terrestrial, satellite and cable. The communication channel typically introduces some noise, and provision for error correction is made in the channel coder to compensate for this channel noise. For detailed discussion on digital modulation and error-correction coding, the interested readers are referred to references [1-4],[1-5],[1-6]. Some basics of modulation and channel coding for video transmission are also provided in Appendix A.

At the head-end and end-user receivers, the received signal is demodulated and transformed back into a string of bits by a channel decoder. The

uncorrectable errors are marked (or say indicated) in the reconstructed packets. Certain video and audio packets can be replaced in the head-end devices by packets from coded ***local video and audio programs,*** e.g. local commercials. The recovered packets are de-scrambled and de-multiplexed with the timing information into separate video and audio streams. The video decoder reconstructs the video for human viewing by using the extracted decoding and presentation time while the audio decoder plays the audio simultaneously. At the receiver, the coded video and audio packets of any given program can be ***randomly accessed*** as program tuning and program switching.

Fig. 1.2 presents another type of DTV service that uses an Asynchronous Transfer Mode (ATM) network. The physical transmission medium include Digital Subscriber Line (DSL) and Optical Fiber. Most parts of the transmission process in this example are similar to examples given in Fig. 1.1. One key feature is that video and audio packets have to be able to indicate packet loss caused by the network. In this case, packet loss is indicated by ***packet counter*** value that is carried in the packets.

The examples discussed above are MPEG-enabled digital video broadcasting systems. MPEG is an ISO/IEC working group whose mandate is to generate standards for digital video and audio compression. Its main goal is to specify the coded bit streams, transporting mechanisms and decoding tool set for digital video and audio. Three video-compression standards have been ratified and they are named MPEG-1 [1-7][1-8][1-9], MPEG-2 [1-10][1-11][1-12], and MPEG-4 [1-13][1-14][1-15].

Distribution networks such as terrestrial, direct-broadcasting satellite and cable television services have exploited the potential of the MPEG standards of digital compression to increase services and lower costs. The standard that used in the broadcast DTV services is MPEG-2. These services can deliver coded video at resolution of ITU-R 601 [1-17] interlaced video (e.g. a size 704 x 480 for NTSC and 704 x 576 for PAL). These services have also been extended to higher resolution and bit rate for the market of High Definition Television (HDTV). For example, it can process video sequences with sampling dimensions at 1920 x 1080 x 30Hz and coded video bit-rates around 19Mbit/s.

The presence of MPEG-1, MPEG-2 and MPEG-4 standards gives the opportunity for a system designer to pick the compression technology that is the best for their particular application. The advantages of both MPEG video compression standards include significant overall saving on system costs, higher quality, and greater programming choices.

Next, consider the MPEG-1 system as a simple example of video transport. The MPEG-1 video standard (ISO/IEC 11172-2) [1-18] specifies a coded representation that can be used for compressing video sequences to bit-rate that was optimized around 1.5 Mbit/s. It was developed to operate primarily from storage media offering a continuous transfer rate of about 1.5 Mbit/s. Nevertheless it is also widely used in many applications. The MPEG-1 system standard (ISO/IEC 11172-1) [1-17] addresses the problem of ***multiplexing*** one or more data streams from the video and audio parts of the MPEG-1 standard with ***timing information*** to form a single stream as in Fig. 1.3 below. This is an important function because, once combined into a single stream, the data are in a form well suited to digital storage or transmission. Thus, the system part of the standard gives the integration of the audio and video streams with the proper time stamping to allow ***synchronization*** of coded bitstreams.

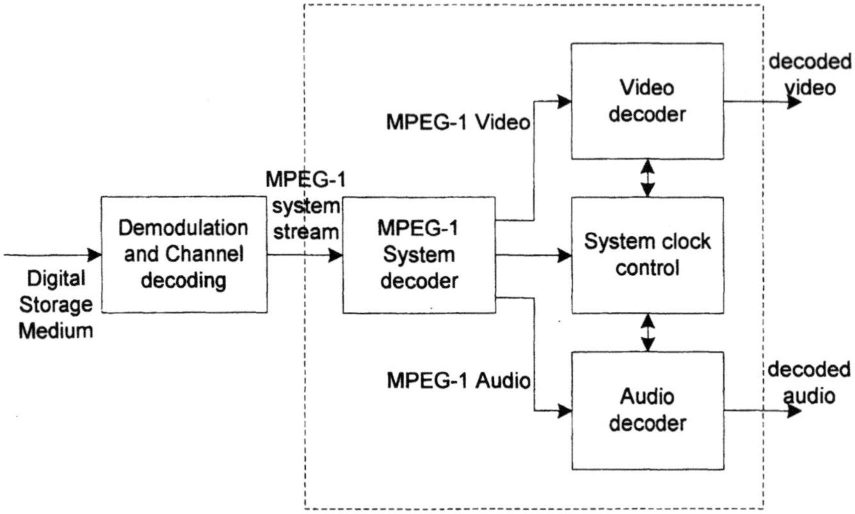

Figure 1.3 MPEG-1 decoding systems

The above examples have clearly described the functional objectives of video transport systems. These objectives can be summarized as follows:

(1) To provide a mechanism for packetizing video data with functionalities such as packet synchronization and identification, error handling, conditional access, random entry into the compressed bit stream and

synchronization of the decoding and presentation process for the applications running at a receiver,

(2) To schedule and multiplex the packetized data from multiple programs for transmission,

(3) To specify protocols for triggering functional responses in the transport decoder, and

(4) To ensure the video bit stream level interoperability between communication systems.

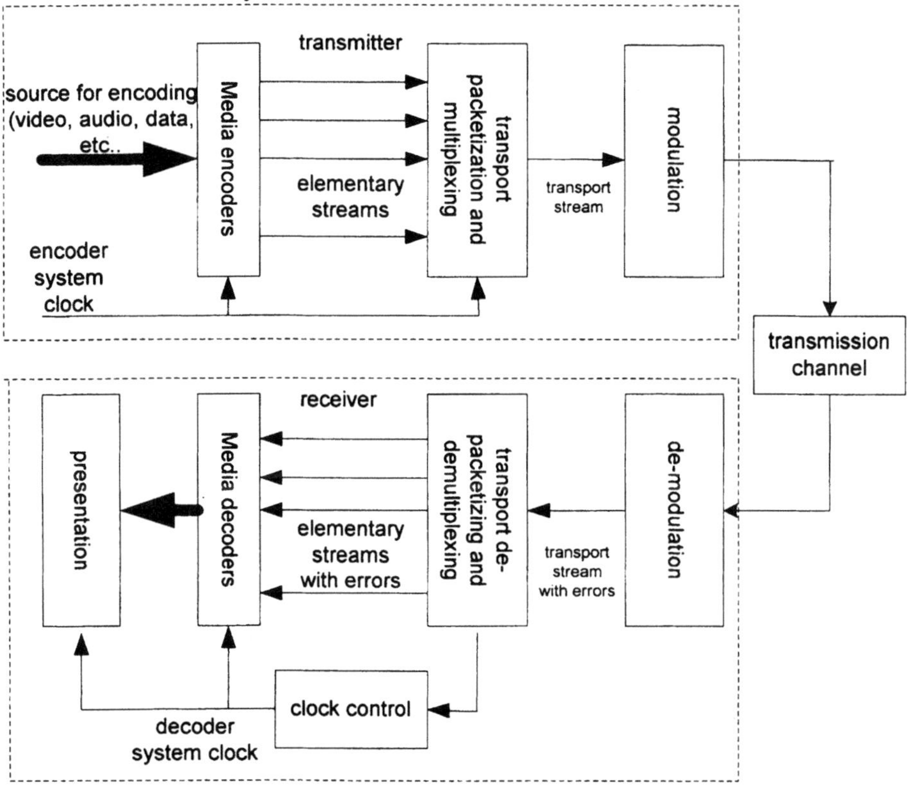

Figure 1.4 A typical transmitter-receiver pair

Fig. 1.4 illustrates the organization of a typical transmitter-receiver pair and the location of the transport subsystem in the overall system. The transport resides between the media data (e.g. audio or video) encoding/decoding function and the transmission subsystems. The encoder transport subsystem is responsible for formatting the encoded bits and multiplexing the different components of the program for transmission. At the receiver, it is responsible for recovering the bit streams for the individual application decoders and for the corresponding error signaling. The transport subsystem also incorporates other functionality related to identification of applications and synchronization of the receiver. This text will discuss in great details about issues in design of these functions. In the following sections of this chapter, an overview of functionality of digital video transport is provided.

1.3 Fixed Length Packet vs. Variable Length Packet

In general there are two approaches for multiplexing elementary streams from multiple applications on to a single channel. One approach is based on the use of fixed length transport packets and the other on variable length transport packets. Both approaches have been used in the MPEG-2 systems standard [1-10]. In MPEG-2 systems, the stream that consists of the fixed length transport packets is called a transport stream (TS) while the stream that consists of variable length packets is called a program stream (PS). In this text, bit streams generated by video and audio compression engines are called elementary streams (ES). As illustrated in Fig. 1.5, the video and audio streams in both TS and PS cases go through an initial stage of packetization, which results in variable length packets called packetized elementary stream (PES). The process of generating the transmitted bit streams for the two approaches is shown to involve a difference in processing at the final multiplexing stage.

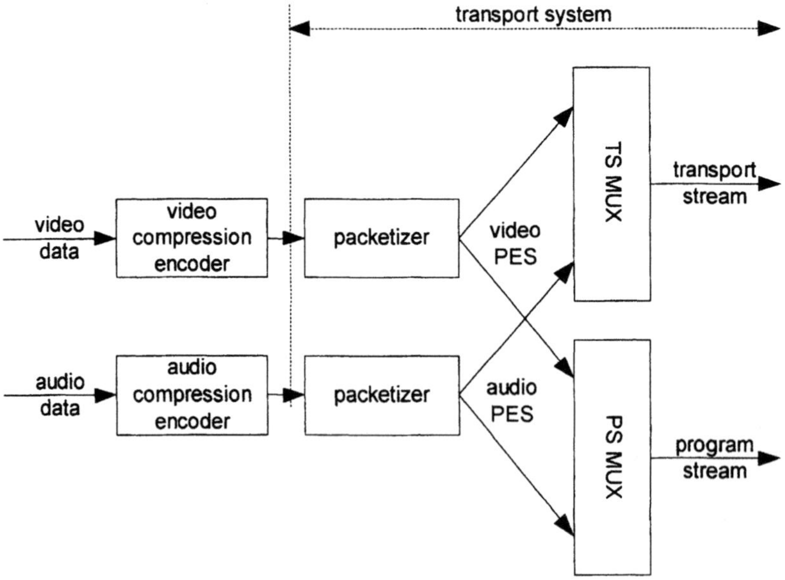

Figure 1.5 Two packetization and multiplexing schemes

Examples of bit streams for the both program and transport stream approaches are given in Fig. 1.6 to clarify their difference. In the TS approach shown by Fig. 1.6a, each PES packet of video or audio stream occupies a variable number of transport packets, and data from video and audio bit streams are generally interleaved with each other at the final transmitted stream, with identification of each elementary bit stream being facilitated by data in the transport headers. In the PS approach shown by Fig. 1.6b, PES packets of video or audio bit stream are multiplexed by transmitting the bits for the complete PES packets in sequence, thus resulting in a sequence of variable length packets on the channel.

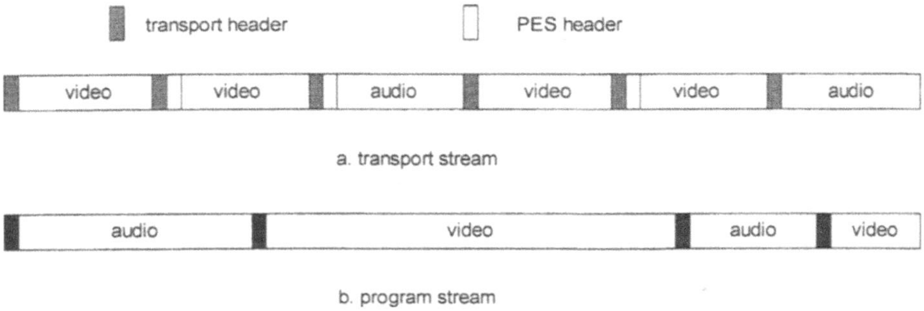

Figure 1.6 Transport and program streams

These two multiplexing approaches are motivated by different application scenarios. Transport streams are defined for environments where errors and data loss events are likely, including storage applications and transmission on noisy channels, e.g. satellite and cable DTV systems. Program streams on the other hand are designed for relatively error-free media, e.g. DVD-ROMs. Errors or loss of data within PES packets can be potentially result in complete loss of synchronization in the decoding process in this case.

In general, the fixed length packetization approach offers a great deal of flexibility and some additional advantages when attempting to multiplex data related to multiple applications on a single bit stream. These are described in some detail below.

Flexible Channel Capacity Allocation: While digital transport systems are generally described as flexible, the use of fixed length packets offers complete flexibility to allocate channel capacity among video, audio and auxiliary data services. The use of a packet-identification word in the packet header as a means of bit stream identification makes it possible to have a mix of video, audio and auxiliary data that is flexible and needs not be specified in advance. The entire channel capacity can be reallocated in bursts for data delivery. This capability can be used in various multimedia services.

Bandwidth Scalability: The fixed-length packet format is scalable in the sense that availability of a larger bandwidth may also be exploited by adding more elementary bit streams at the input of the multiplexer, or even multiplexing these elementary bit streams at the second multiplexing stage with the original bit stream. This is a critical feature for network distribution, and also services interoperability with cable or satellite transmission capability to deliver a higher data rate for a given bandwidth.

Service Extensibility: This is a very important factor that needs to be considered for future services that we cannot anticipate today. The fixed-length packet format allows new elementary bit streams being handled without hardware modification, by assigning new packet identification words at the transmitter to new packets and filtering on these new packets in the bit stream at the receiver. Backward compatibility is assured when new bit streams are introduced into the transport system since existing decoders will automatically ignore packets with new identification words.

Transmission Robustness: This is another advantage of the fixed length packetization approach. The fixed-length packet provides better and simpler ways for handling errors that are introduced in transmission. Error correction and detection processing may be synchronized to the packet structure so that one only needs to deal at the decoder with unit of packets when handling data loss due to transmission impairments. After detecting errors during transmission, one can recover the coded bit stream from the first uncorrupted packet. Recovery of synchronization within each application is also added by the transport packet header information. Without this approach, recovery of synchronization in the bit streams would have been completely dependent on the properties of each elementary bit stream.

Cost effective implementation: A fixed-length packet based transport system enables simple decoder bit stream de-multiplex architectures, suitable for high-speed implementations. The decoder does not need detailed knowledge of the multiplexing strategy or parameters of the coded source to extract individual elementary bit streams at the de-multiplexer. What the receiver needs to know is the identity of the packet, which is transmitted in each packet header at fixed and known locations in the bit stream. The most important information is the timing information for elementary stream level and packet level synchronization.

In this book, we focus on discussion of the data transport mechanism that is based on the use of fixed length packets.

1.4 The Packetization Approach and Functionality

The fixed length packet usually has a format shown in Fig. 1.7 [1-3]. The so-called "link" header contains fields for packet synchronization and identification, error indication, and conditional access. The adaptation header carries synchronization and timing information for decoding and presentation process. It can also provide indicators for random access points of compressed bit streams and for "local" program insertion. The payload could be any multimedia data including compressed video and audio streams.

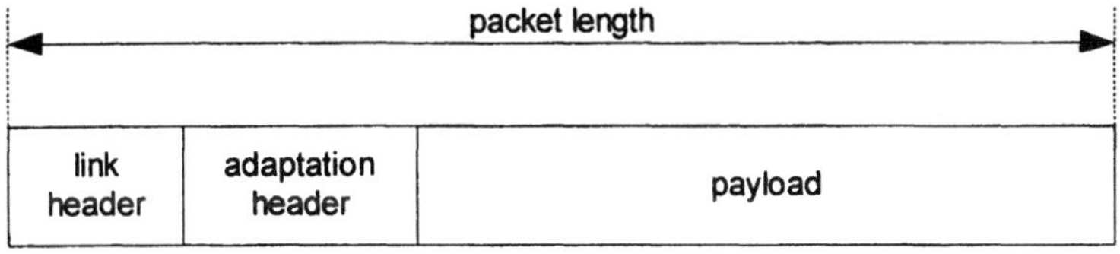

Figure 1.7 The fixed-length packet format

The MPEG-2 transport packet consists of 188 bytes. The choice of this packet size is motivated by a few key factors at the time. The packets need to be large enough so that the overhead of the transport headers does not become a significant portion of the total data being carried. The packet size should not be too large that the probability of packet error becomes significant under standard operating conditions (due to inefficient error correction). It is also desirable to have packet lengths in tune with the block size of typical, block oriented, error correction approaches, so that packets may be synchronized to error correction blocks, and the physical layer of the system can aid the packet level synchronization process in the decoder. Another motive for the particular packet length selection is interoperability with the ATM packet. The general philosophy is to transmit a single MPEG-2 transport packet in four ATM packets.

The contents of each packet and the nature of this data are identified by the packet headers. The packet leader structure is layered and may be described as a combination of a fixed length "link" layer and a variable length adaptation layer. Each layer serves a different functionality similar to the link and transport layer functions in the Open System Interconnection Reference Model [1-3]. This link and adaptation level functionality is directly used for the various transmission networks such as satellite and cable digital television networks.

1.4.1 The "link" layer header

The "link" layer header field can support the following important functions.

Packet synchronization is usually enabled by the synchronization word at beginning of a packet. This word has the same fixed, pre-assigned, value for all packets. For example, the synchronization word in MPEG-2 transport stream is the first byte in a packet and has a pre-assigned value of 0x47. In some implementations of decoders the packet synchronization function is done at the physical layer of the communication link that precedes the packet de-multiplexing stage. In this case, the synchronization word field may be used for verification of packet synchronization function. In other decoder implementations this word may be used as the primary source of information for establishing packet synchronizations.

Packet identification field is needed in each packet. This is usually called the Packet ID (PID) in MPEG-2. It provides the mechanism for multiplexing and de-multiplexing bit streams, by enabling identification of packets belonging to a particular elementary or control bit stream. Since the location of the PID field in a packet header is always fixed, extraction of the packets corresponding to a particular elementary bit stream is very simply achieved once packet synchronization is established by filtering packets based on PIDs. Some simple filter and de-multiplexing designs can be implemented for fixed length packets. These implementations are suitable for high-speed transmission systems.

Error Handling fields are used to assist the error detection process in the decoder. Error detection is enabled at the packet layer in the decoder through the use of the packet error flag and packet counter. In MPEG-2, these two fields are the transport_packet_error_indicator field (1-bit) and the continuity_counter field (4-bits). When uncorrectable errors are detected by the error-correction subsystem, the transport_packet_error_indicator fields in the corresponding packets are marked. At the transmitter end, the value in the continuity_counter field cycles from 0 through 15 for all packets with the same PID that carry a data payload. At the receiver end, under normal conditions, the reception of packets in a PID stream with a discontinuity in the continuity_counter value indicates that data has been lost in transmission. The transport processor at the decoder then signals the decoder for the particular elementary stream about the loss of data. Because certain information (such as headers, time stamps, and program maps) is very important to the smooth and continuous operation of a system, the transport system has a means of increasing the robustness of this information to channel errors by providing a mechanism for the encoder to duplicate packets. Those packets that contain important information will be duplicated at the encoder. At the decoder, the duplicate packets are either used if the original packet was in error or are dropped.

Access control is defined as protection against unauthorized use of resources, including protection against the use of resources in an unauthorized manner. Digital video transmission systems have to provide access control facilities to be economically viable. The sooner these facilities are taken into account in the definition, specification and implementation of the systems, the earlier their deployments are. A complete access control system usually includes three main functions: the scrambling/de-scrambling function, the entitlement control function and the entitlement management function.

The scrambling/de-scrambling function aims at making the program incomprehensible for unauthorized receivers. Conditional Access indication field is provided in the "link" layer header. The transport format allows for scrambling of data in the packets. Scrambling can be applied separately to each elementary bit-stream. De-scrambling is achieved by the receiver withholding a secret key used for a scrambling algorithm. Usually, the transport packet specifies the de-scrambling approach to be used but does not specify the de-scrambling key and how it is obtained at the decoder.

The entitlement control function provides the conditions required to access a scrambled program together with the encrypted secret parameters enabling the de-scrambling for the authorized receivers. These data are broadcasted as conditional access messages, called Entitlement Control Messages (ECMs), which carries an encrypted form of the keys or a means to recover the keys, together with access parameters, i.e. a identification of the service and of the conditions required for accessing this service.

The entitlement management function consists in distributing the entitlements to the receivers. There are several kinds of entitlements matching the different means to "buy" a TV program. These entitlement data are also broadcasted as conditional access messages, called entitlement management messages (EMMs), used to convey entitlements or keys to users, or to invalidate or delete entitlements or keys.

The key must be delivered to the decoder within a time interval of its usefulness. Both ECM and EMM can be carried at several locations within the transport stream. For example, two likely locations would be (1) as a separate private stream with it's own PID, or (2) a private field within an adaptation header carried by the PID of the signal being scrambled. The security of the conditional access system is ensured by encrypting the de-scrambling key when sending it to the receiver, and by updating the key frequently. Usually, the key encryption, transmission, and decryption approaches could differ in different ATV delivery systems. There is not a system-imposed limit on the number of keys that can be used and the rate at which these may be changed. The only requirement for conditional access in a receiver is to have an interface from the decryption approach and technology is itself not a part of specification of transport packet.

Information in the link header of a transport packet describes if the payload in the packet is scrambled and if so, flags the key to be used for de-scrambling. The header information in a packet is always transmitted in the clear, i.e., unscrambled.

In MPEG-2 transport system, the mechanism for scrambling functions are provided at two levels, within the PES packet structure and at the transport layer. Scrambling at the PES packet layer is primarily useful in the program stream, where there is no protocol layer similar to the transport to enable this function.

1.4.2 The Adaptation Layer

The adaptation header in the transport packet is usually a variable length field. Its presence is conditional to some flags in the link header. The functionality of these headers is basically related to the decoding of the elementary bit stream that is extracted using the link level functions. Some of the functions of the adaptation layer are described next.

Random access is the process of beginning to read and decoded the coded bit stream at an arbitrary point. Random access points, as random entry points into the compressed bit streams, can be indicated in the adaptation layer of the packet. For video and audio, such entry points are necessary to support functions such as program tuning and program switching. Random entry into an application is possible only if the coding for the elementary bit stream for the application supports this functionality directly. For example, a compressed video bit stream supports random entry through the concept of Intra (or I-) frames that are coded without any prediction between adjacent pictures, and which can therefore be decoded without any information from prior pictures. The beginning of the video sequence header information preceding data for an I-frame could serve as a random entry point into a video elementary bit stream. In MPEG-2 system, random entry points, in general, should also coincide with the start of PES packets where they are used, e.g., for video and audio. The support for random entry at the transport layer comes from a flag in the adaptation header of the packet that indicates whether the packet contains a random access point for the elementary bit stream. In addition, the data payload of packets that are random access points also start with the data that forms the random access points into the elementary bit stream itself. This approach allows the discarding of packets directly at the transport layer when switching channels and searching for a resynchronization point in the transport bit stream, and also simplifies the search for the random access point in the elementary bit stream once transport level resynchronization is achieved. One objective is to have random entry points into the programs as frequently as possible, to enable rapid channel switching.

Splicing system supports the concatenation, performed on the transport packet level, of two different elementary streams. The spliced stream might results in discontinuities in time-base, continuity counter, control bit streams, and video decoding. Splicing point is important for inserting local programming, e.g. commercials, into a bit stream at a broadcast head-end. In general, there are only certain fixed points in the elementary bit streams at which program insertion is allowed. The local insertion points has to be a random entry point but not all random entry points are suitable for program insertion. Local program insertion also always takes place at the transport packet layer, i.e., the data stream splice points are packet aligned. Implementation of the program insertion process by the broadcaster is aided by the use of a counter field in the adaptation header that indicates ahead of time the number of packets to countdown until the packet after which splicing and local program insertion is possible.

Video synchronization is often required even if the video signals are transmitted through synchronous digital networks because video terminals generally work independently of the network clock. In the case of packet transmission, packet jitter caused by packet multiplexing also has to be considered. This implies that synchronization in packet transmission may become more different than with synchronous digital transmission. Hence, video synchronization functions that consider these conditions should be introduced into video transport systems.

Synchronization and timing information can be carried in the adaptation layer in terms of time-stamps such as sampled system clock values. A discussion on synchronization and timing is given in the next section.

1.5 Buffer, Timing and Synchronization

Uncompressed video is constant rate by nature and is transmitted over constant-rate channels, e.g. analog TV signal over cable broadcast network. For transmission of compressed digital video, since most video compression algorithms use variable length codes, an encoder buffer is necessary to translate the variable rate output by the compression engine into the constant-rate channel. A similar buffer is also necessary at the receiver to convert the constant channel bit rate into a variable bit rate. It will be shown in Chapter 3 that for a constant-rate channel, it is possible to prevent the decoder buffer from over-flowing or under-flowing simply by ensuring that the encoder buffer never underflows or overflows.

In general case, compressed video can also be transmitted over variable-rate channels, e.g. multiplexed transport channels and broadband IP networks. These networks are able to support variable bit rates by partitioning video data into a sequence of packets and inputting them to the network asynchronously. In another words, these networks may allow video to be transmitted on a channel with variable rate. For a variable-rate channel, additional constraints must be imposed on the encoding rate, the channel rate, or both.

Synchronization and timing recovery process specified in the transport system involves the sampling of the analog signals, encoding, encoder buffering, transmission, reception, decoder buffering, decoding, and presentation of digital audio and video in combination.

Synchronization of the decoding and presentation process for the applications running at a receiver is a particularly important aspect of real time digital data delivery systems. Since received packets are processed at a particular rate (to match the rate at which it is generated and transmitted), loss of synchronization leads to either buffer overflow or underflow at the decoder, and as a consequence, loss of presentation/display synchronization. The problems in dealing with this issue for a digital compressed bit stream are different from those for analog NTSC or PAL. In NTSC or PAL, information is transmitted for the pictures in a synchronization manner, so that one can derive a clock directly from the picture synch. In a digital compressed system the amount of data generated for each picture is variable (based on the picture coding approach and complexity), and timing cannot be derived directly from the start of picture data. Indeed, there is really no natural concept of synch pulses (that one is familiar with in NTSC or PAL) in a digital bit stream.

One solution to this issue in a transport system is to transmit timing information in the header of selected packets, to serve as a reference for timing comparison at the decoder. This can be done by transmitting a sample the system clock in the specified field, which indicates the expected time at the completion of the reading of that field from the bit stream at the transport decoder. The phase of the local system clock running at the decoder is compared to the sampled value in the bit stream at the instant at which it is obtained, to determine whether the decoding process is synchronized. In general, the sampled clock value in the bit stream does not directly change the phase of the local clock but only serves as an input to adjust the clock rate. Exceptions are during the time base changes, e.g. channel change. The audio and video sample clocks in the decoder system are locked to the system clock derived from the sampled clock values. This simplifies the receiver implementation in terms of the number of local oscillators required to drive the complete decoding process, and has other advantages such as rapid synch acquisition. In this book, both principle and implementation of the timing recovery process are discussed.

MPEG-2 transport system specification provides a timing model in which all digitized pictures and audio samples that enter the video compression engines are presented exactly once each, after a constant end-to-end delay, at the output of the decompression engines. The sample rates, i.e. the video frame rate and the audio sample rate, are precisely the same at the inputs of the compression engines as they are at the outputs of the decompression engines. This timing model is diagrammed in Fig. 1.8.

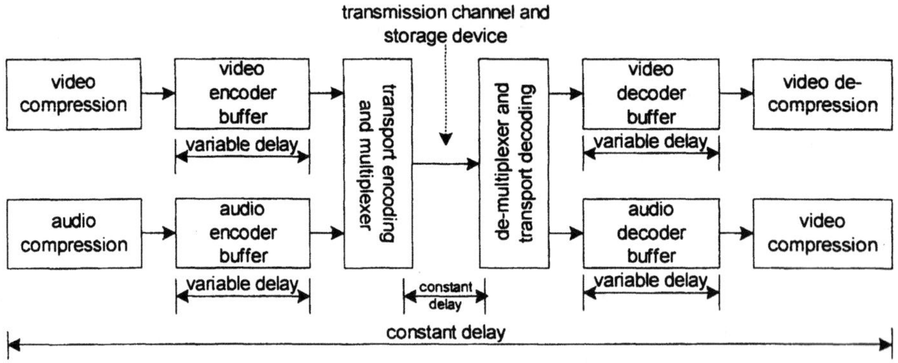

Figure 1.8 The timing model in MPEG-2 system

As shown in Fig. 1.8, the delay from the input to the compression engine to the output or presentation from the decompression engine is constant in this model while the delay through each of the encoder and decoder buffers is variable. Not only is the delay through each of these buffers variable within the path of one elementary stream, the individual buffer delays in the video and audio paths differ as well. Therefore the relative location of coded bits representing audio or video in the combined stream does not indicate synchronization information. The relative location of coded audio and video is constrained only by the System Target Decoder (STD) model such that the decoder buffers must behave properly; therefore coded audio and video that represent sound and pictures that are to be presented simultaneously may be separated in time within the coded bit stream by as much as one second, which is the maximum decoder buffer delay that is allowed in the STD model.

The audio and video sample rates at the inputs of compression engines are significantly different from one another, and may or may not have an exact and fixed relationship to one another. The duration of an audio presentation unit is generally not the same as the duration of a video picture.

In MPEG-2 system, there is a single, common system clock in the compression engines for a program, and this clock is used to create timestamps that indicate the presentation and decoding timing of audio and video, as well as to create timestamps that indicate the instantaneous values of the system clock itself at sampled intervals. The timestamps that indicate the presentation time of audio and video are called Presentation Time Stamps (PTS). Those that indicate the decoding time are called Decoding Timestamps (DTS), and those that indicate the value of the system clock are called the System Clock Reference (SCR) in Program Streams and the Program Clock Reference (PCR) in Transport Streams. It is the presence of this common system clock in the compression engines, the timestamps that are created from it, and the recreation of the clock in the decompression engines and the correct use of the timestamps that provide the facility to synchronize properly the operation of the decoding.

Since the end-to-end delay through the entire system is constant, the audio and video presentations are precisely synchronized. The construction of bit streams is constrained such that, when they are decompressed with the appropriately sized decoder buffers, those buffers are guaranteed neither overflow nor underflow.

In order for the decompression engine to incur the precise amount of delay that ensures the entire end-to-end delay to be constant, it is necessary for the decompression engine to have a system clock whose frequency of operation and absolute instantaneous value match those of the compression engine. The information necessary to convey the system clock can be encoded in the transport bit stream.

If the clock frequency of the decompression engine matches exactly that of the corresponding compression engine, then the decoding and presentation of video and audio will automatically have the same rate as those at the encoding process, and the end-to-end delay will be constant. With matched encoding and decoding clock frequencies, any correct value of the sampled encoding system clock, e.g. the correct PCR in MPEG-2 transport streams, can be used to set the instantaneous value of the decoding system clock, and from that time on the decoding system clock will match that of the encoder without the need for further adjustment. However, in practice, the free-running decoding system clock frequency will not match the encoding system clock frequency that is sampled and transmitted in the stream. The decoding system clock can be made to slave its timing to the encoding process by using the received encoding system clock samples. The typical method of slaving the decoding clock to the received data stream is via a phase-locked loop (PLL).

Transport systems that are designed in accordance with the MPEG-2 type of system timing model such that decompression engines present audio samples and video pictures exactly once at a constant rate, and such that decoder buffers behave as in the model, are referred to in this book as precisely timed systems. In some applications, video transport systems are not required to present audio and video in accordance with the MPEG-2 type of system timing model. For example, the Internet video transport systems usually do not have constant delay, or equivalently do not present each picture or audio sample exactly once. In such systems, the synchronization between presented audio and video may not be precise, and the behavior of the decoder buffers may not follow any model. Nevertheless, it is important to avoid overflow at the decoder buffers, as overflow causes a loss of data that may have significant effects on the resulting decoding process.

Buffer constraints on compressed digital video are discussed in greater detail in Chapter 3 while design issues related to timing and synchronization are studied in Chapters 4, 5 and 6.

1.6 Multiplexing Functionality

As described earlier, the overall multiplexing approach can be described as a combination of multiplexing at two different layers. In the first layer, a single-program transport stream is formed by multiplexing one or more elementary bit streams at the transport layer, and in the second layer the multiple program transport streams are combined (using asynchronous packet multiplexing) to form the overall system. The functional layer in the system that contains both this program and system level information that is going to be described is called the Program Specific Information (PSI).

A typical single-program transport bit stream consists of packetized elementary bit streams (or just elementary stream) that share a common system clock (sometimes called the time-base), and a control bit stream that describes the program. Each elementary bit stream, and the control bit stream (also called the elementary stream map in Figure 1.9), are identified by their unique PIDs in the link header field. The organization of the multiplex function is illustrated in Fig. 1.9. The control bit stream contains the program_map_table that describes the elementary stream map. The program_map_table includes information about the PIDs of the transport streams that make up the program, the identification of the applications that are being transmitted on these bit streams, the relationship between these bit streams, etc.. The details of the program_map_table syntax and the functionality of each syntax element are given in a later section. The identification of a bit-stream carrying a program_map_table is done at the system layers to be described next.

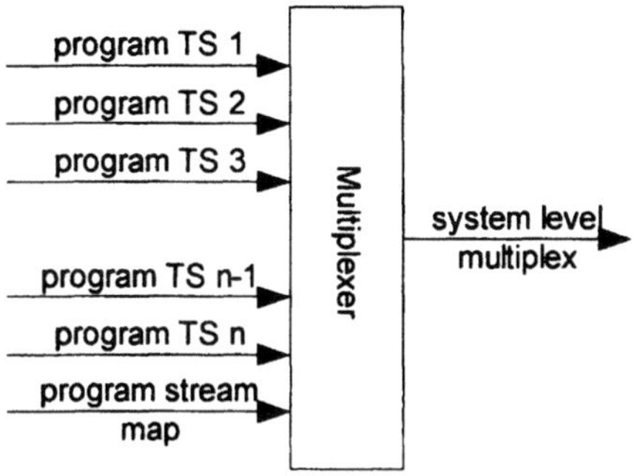

Figure 1.9 A functional diagram of stream multiplexer

In general, the transport format allows a program to be comprised of a large number of elementary bit streams, with no restriction on the types of applications required within a program. A transport bit stream does not need to contain compressed video or audio bit streams, or, for example, it could contain multiple audio bit streams for a given video bit stream. The data applications that can be carried are flexible, the only constraint being that there should be an appropriate stream_type ID assignment for recognition of the application corresponding to the bit stream in the transport decoder. Usually, the process of identifying a program and its contents takes place in two stages: first one uses the program_association_table in the PID=0 bit stream to identify the PID of the bit stream carrying the program_map_table for the program, in the next stage one obtains the PIDs of the elementary bit streams that make up the program from the appropriate program_map_table. Once this step is completed the filters at a demultiplexer can be set to receive the transport bit streams that correspond to the program of interest.

The system layer of multiplexing is illustrated in Fig. 1.10. Note that during the process of system level multiplexing, there is the possibility of PIDs on different program streams being identical at the input. This poses a problem since PIDs for different bit streams need to be unique. A solution to this program lies at the multiplexing stage, where some of the PIDs could be modified just before the multiplex operation. The changes have to be recorded in both the program_association_table and the program_map_table. Hardware implementation of the PID reassignment function in real time is helped by the fact that this process is synchronous at the packet clock rate. The other approach, of course, is to make sure up front that the PIDs being used in the programs that make up the system are unique. This is not always possible with stored bit streams.

Since the architecture of a transport bit stream is usually scalable, multiple system level bit streams can be multiplexed together on a higher bandwidth channel by extracting the program_association_tables from each system multiplexed bit stream and reconstructing a new PID=0 bit stream. Note that PIDs may have to be reassigned in this case.

In the above descriptions of the higher level multiplexing functionality no mention is made of the functioning of the multiplexer and multiplexing policy that should be used. In general, the transport demultiplexer will function on any transport bit stream regardless of the multiplexing algorithm used. The multiplexing algorithms will be discussed in Chapter 8.

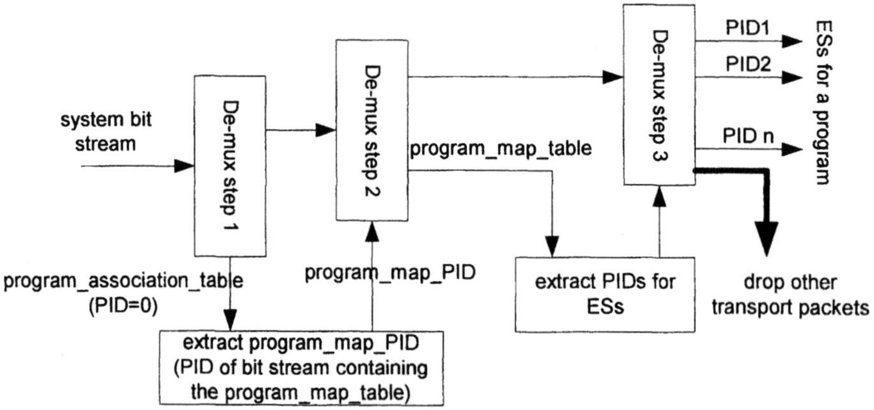

Figure 1.10 The process of extracting elementary bit streams

Fig. 1.10 illustrates the entire process of extracting elementary bit streams for a program at a receiver. It also services as one possible implementation approach. In practice the same demultiplexer hardware could be used to extract both the program_association_table and the program_map_table control bitsteams. This also represents the minimum functionality required at the transport layer to extract any application bit stream including video, audio and other multimedia data streams.

Once the packets are obtained from each elementary bit stream in the program, further processing stages of obtaining the random access points for each video and audio elementary bit stream, decoder system clock synchronization, presentation (or decoding) synchronization, etc.., need to take place before the receiver decoding process reaches normal operating conditions for receiving a program.

It is important to clarify here that the layered approach to define the multiplexing function does not necessarily imply that program and system multiplexing should always be implemented in separate stages. A hardware implementation that includes both the program and system level multiplexing within a single multiplexer stage is a common practice.

Chapters 8 and 9 of this book cover the topics of multiplexing technologies for video transporting systems.

1.7 Interoperability, Transcoding and Re-multiplexing

In this book, we focus on the data transport mechanism that is based on the use of fixed length packets that are identified by headers. Each header identifies a particular application bit stream, e.g. a video or audio elementary bit stream, that forms the payload of the packets. Applications supported also include data program and system control information, etc.. As indicated earlier, the elementary bit streams for video and audio are themselves been wrapped in a variable length packet structure called the packet elementary stream (PES) before transport processing. The PES provides functionality for identification, and synchronization of decoding and presentation of the individual application.

Elementary bit streams sharing a common system clock are multiplexed, along with a control data stream, into programs. These programs and an overall system control data stream are then asynchronously multiplexed to form a multiplexed system. Fig. 1.11 summarizes a layered transport data flow with its functionality.

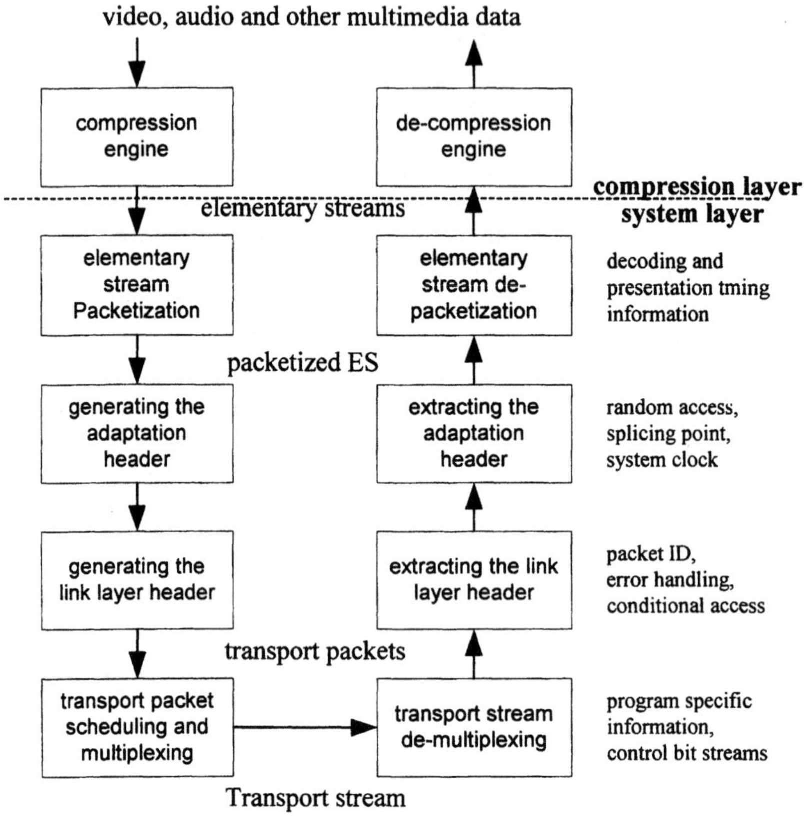

Figure 1.11 A layered transport architecture.

Due to the variety of different networks comprising the present communication infrastructure, a connection from the video source to the end user may be established through links of different characteristics and bandwidth.

The question has been raised frequently about the bit stream level interoperability of the transport system. There are two sides to this issue. One is whether a transport bit stream for one system can be carried on other communication systems, and the other is the ability of the transport system to carry bit streams generated from other communication systems.

The first aspect of transmitting transport bit streams in different communication systems will be addressed to some extent in later chapters. In short, there should be nothing that prevents the transmission of a well-specified transport bit stream as the payload on a different transmission system. It may be simpler to achieve this functionality in certain systems, e.g. Cable Television system (CATV), direct broadcasting system (DBS), ATM, etc., than in others, e.g., data networks based on protocols such as Real Time Protocol (RTP), etc..

The other aspect is of transmitting other bit streams within a transport system. This makes more sense for bit streams linked to TV broadcast applications, e.g. CATV, DBS, etc.., but is also possible for other types of bit streams. This function is achieved by transmitting these other bit streams as the payload of identifiable transport packets. The only requirement is to have the general nature of these bit streams recognized within the specified transport system content.

In order to transmit the compressed video over the networks with different characteristics and bandwidth, video transport packets have to be able to adapt the changes in the video elementary stream. In the case where only one user is connected to the source, or independent transmission paths exist for different users, the bandwidth required by the compressed video should be adjusted by the source in order to match the available bandwidth of the most stringent link used in the connection. For uncompressed video, this can be achieved in video encoding systems by adjusting coding parameters, such as quantizition steps, whereas for pre-compressed video, such a task is performed by applying, so called, video transcoders [1-18], [1-19].

In the case where several users are simultaneously connected to the source and receiving the same coded video, as happen in video on demand (VoD) services, CATV services and Internet video, the existence of links with different capacities poses a serious problem. In order to deliver the same compressed video to all users, the source has to comply with the sub-network that has the lowest available capacity. This unfairly penalizes those users that have wider bandwidth in their own access links. By using transcoders in communication links, this problem can be resolved. For a video network with transcoders in its subnets, one can ensure that users receiving lower quality video are those having lower bandwidth in their transmission paths. An example of this scenario is in CATV services where a satellite link is used to transmit compressed video from the source to a ground station, which in turn distributes the received video to several destinations through networks of different capacity. Ground stations, such as cable head-ends, can re-assemble programs from different video sources. Some programs from broadcast television and others from video servers are re-multiplexed for transmission. A re-multiplexer is a device that receives one or more multi-program transport streams and retains a subset of the input programs, and outputs the retained programs in such a manner that the timing and buffer constraints on output streams are satisfied. In order to ensure that the re-assembled programs can match the available bandwidth, video transcoders can be used along with the re-multiplexer to allow bit-rate reduction of the compressed video.

Buffer analysis and management of transcoding systems are discussed in Chapter 7 and the re-multiplexing techniques are introduced in Chapters 8 and 9.

Bibliography

For books and articles devoted to video transporting systems:

[1-1] B. G. Haskell, A. Puri, and A. N. Netravali, Digital Video: An Introduction to MPEG-2, New York: Chapman & Hall, 1997.
[1-2] Ralf Schafer and Thomas Sikora, "Digital video coding standards and their role in video communications", Proceeding of IEEE, Vol. 83, No. 6, pp.907-924, June 1995.
[1-3] A54, Guide to the use of the ATSC digital television standard, Advanced Television Systems Committee, Oct. 19, 1995.
[1-4] Irving S. Reed and Xuemin Chen, Error-Control Coding for Data Networks, 2nd print, Kluwer Academic Publishers, Boston, 2001.
[1-5] Irving S. Reed and Xuemin Chen, article Channel Coding, Networking issue of Encyclopedia of Electrical and Electronic Engineering, John Wiley & Sons, Inc. New York, Feb., 1999.
[1-6] Jerry Whitaker, DTV Handbook, 3rd Edition, McGraw-Hill, New York, 2001.
[1-7] ISO/IEC 11172-1:1993, Information technology – Coding of moving pictures and associated audio for digital storage media at up to about 1,5 Mbit/s – Part 1: Systems.

[1-8] ISO/IEC 11172-2:1993, Information technology — Coding of moving pictures and associated audio for digital storage media at up to about 1,5 Mbit/s — Part 2: Video.

[1-9] ISO/IEC 11172-3:1993, Information technology — Coding of moving pictures and associated audio for digital storage media at up to about 1,5 Mbit/s — Part 3: Audio.

[1-10] ITU-T Recommendation H.222.0(1995) | ISO/IEC 13818-1:1996, Information technology — Generic coding of moving pictures and associated audio information: Systems.

[1-11] ITU-T Recommendation H.262(1995) | ISO/IEC 13818-2:1996, Information technology — Generic coding of moving pictures and associated audio information: Video.

[1-12] ISO/IEC 13818-3:1996, Information technology — Generic coding of moving pictures and associated audio information — Part 3: Audio.

[1-13] ISO/IEC 14496-1:1998, Information Technology - Generic coding of audio-visual objects - Part 1: System.

[1-14] ISO/IEC 14496-2:1998, Information Technology - Generic coding of audio-visual objects - Part 2: Visual.

[1-15] ISO/IEC 14496-3:1998, Information Technology - Generic coding of audio-visual objects - Part 3: Audio.

[1-16] Michael , Data Broadcasting, McGraw-Hill, New York, 2001.

[1-17].J. Watkinson, The Art of Digital Video, Focal Press, Boston, 1990.

[1-18] Xuemin Chen and Fan Ling, "Implementation architectures of a multi-channel MPEG-2 video transcoder using multiple programmable processors", US Patent No. 6275536B1, Aug. 14, 2001.

[1-19] Xuemin Chen, Limin Wang, Ajay Luthra, Robert Eifrig, "Method of architecture for converting MPEG-2 4:2:2-profile bitstreams into main-profile bitstreams", US Patent No. 6259741B1, July 10, 2001.

2 Digital Video Compression Schemes

2.1 Video Compression Technology

Digital video communication is a rapidly evolving field for telecommunication, computer, television and media industries. The progress in this field is supported by the availability of digital transmission channels, digital storage media and efficient digital video coding. Digital video coding often yields better and more efficient representations of video signals. The uncompressed video data often require very high transmission bandwidth and considerable storage capacity. In order to reduce transmission and storage cost, bit rate compression are employed in coding of video signals.

As shown in [2-1]-[2-6] there exist various compression techniques that are in part competitive and in part complementary. Many of these techniques are already applied in industries, while other methods are still undergoing development or are only partly realized. Today and in the near future, the major coding schemes are linear predictive coding, layered coding, and transform coding. The most important image and video compression techniques are:

1. Entropy coding (e.g. run-length coding, Huffman coding and arithmetic coding) [2-9],
2. Source coding (e.g. vector quantization, sub-sampling, and interpolation), transform coding (e.g. Discrete Cosine Transform (DCT) [2-2] and wavelet transform), standardized hybrid coding (e.g. JPEG [2-14], MPEG-1 [2-16], MPEG-2 [2-17], MPEG-4 [2-18], H.261 [2-19], and H.263 [2-20]),

3. Proprietary hybrid-coding techniques (e.g. Intel's Indeo, Microsoft's Window Media Player, Real Networks's Real Video, General Instrument's DigiCipher, IBM's Ultimotion Machine, and Apple's Quick Time, etc.).

The purpose of this chapter is to provide the reader with a basic understanding of the principles and techniques of image and video compression. Various compression schemes, either already in use or yet to be designed, are discussed for transforming signals such as image and video into a compressed digital representation for efficient transmission or storage. Before embarking on this venture, it is appropriate to first introduce and clarify the basic terminology and methods for signal coding and compression.

2.2 Basic Terminology and Methods for Data Coding

The word *signal* originally refers to a continuous time and continuous amplitude waveform, called an *analog signal*. In general sense, people often view a signal as a function of time, where time may be continuous or discrete and where the amplitude or values of the function may be continuous or discrete and may be scalar or vector-valued. Thus by a *signal* we mean a sequence or a waveform whose values at any time is a real number or real vector. In many applications a *signal* also refers to an image which has an amplitude that depends on two spatial coordinates instead of one time variable; or it can also refer to a video (moving images) where the amplitude is a function of two spatial variables and a time variable. The word *data* is sometimes used as a synonym for signal, but more often it refers to a sequence of numbers or more generally, vectors. Thus *data* can often be viewed as discrete time signal. During recent years, however, the word *data* has been increasingly associated in most literature with the discrete or digital case, that is, with discrete time and discrete amplitude, what is called, *digital signal* [2-1].

Physical sources of visual signals such as image and video are analog and continuous time in nature. The first step to convert analog signals to digital form is *sampling*. An analog continuously fluctuating waveform can usually be characterized completely from knowledge of its amplitude values at a countable set of points in time so that one can in effect "throw away" the rest of the signal. It is remarkable that one can discard so much of the waveform and still be able to accurately recover the missing pieces. The intuitive idea is that if one periodically samples data at regularly spaced in time, and the signal does not fluctuate too quickly so that no unexpected wiggles can appear between two consecutive sampling instants, then one can expect to recover the complete waveform by a simple process of interpolation or smoothing, where a smooth curve is drawn that passed through the known amplitude values at the sampling instants.

When watching a movie, one is actually seeing 24 still pictures flashed on the screen every second. Actually, each picture is flashed twice. The movie camera that produced these pictures was actually photographing a scene by taking one still picture every 1/24th of a second. Yet, people have the illusion of seeing continuous motion. In this case, the cinematic process works because human brain is somehow doing the interpolation. This is an example of sampling in action in people's daily lives.

For an electrical waveform, or any other one-dimensional signal, the samples can be carried as amplitudes on a periodic train of narrow pulses. Consider a scalar time function C_b that has a Fourier transform $X(f)$. Assume that there is a finite upper limit on how fast $x(t)$ can wiggle around or vary in time *t*. Specifically, assume that $X(f)=0$ for $|f|\geq W$. That is, the signal has a low-pass spectrum with cutoff frequency *W* Hertz (Hz). To sample this signal, the amplitude is periodically observed at isolated time instants $t = kT$ for *k*=...,-2.-1,0,1,2,.... The *sample rate* is $f_s = 1/T$ and T is the sampling period or sampling interval in seconds.

The idealized case of sampling model is the impulse sampling with a perfect ability to observe isolated amplitude values at the sampling instants *kT*. The effect of such a sampling model is viewed as the process of multiplying the original signal $x(t)$ by a sampling function, $s(t)$, which is the periodic train of impulses $p(t)$ (e.g. Dirac delta functions for ideal case) given by

$$s(t) = T\sum_{k=-\infty}^{\infty} p(t-kT), \tag{2.1}$$

where the amplitude scale is normalized to T so that the average value of $s(t)$ is unity. In the time domain, the effect of this multiplication operation is to generate a new impulse train whose amplitudes are samples of the waveform $x(t)$. Thus,

$$y(t) = x(t)s(t) = T\sum_{k=-\infty}^{\infty} x(t)p(t-kT) \approx T\sum_{k=-\infty}^{\infty} x(kT)p(t-kT). \tag{2.2}$$

Therefore, the signal $y(t)$ contains only the sample values of $x(t)$ and all values in between the sampling instants have been discarded.

The complete recovery of $x(t)$ from the sampled signal $y(t)$ can be achieved if the sampling process satisfies the following fundamental theorem [2-1].

The ***Nyquist Sampling Theorem***: a signal $x(t)$ bandlimited to W (Hz) can be exactly reconstructed from its samples $y(t)$ when it is periodically sampled at a rate $f_s \geq 2W$.

This minimum sampling frequency of $2W$ (Hz) is called the Nyquist frequency or Nyquist rate. If the condition of the sampling theorem is violated, i.e. the sampling rate is less than twice of the maximum frequency component in the spectrum of the signal to be sampled, then the recovered signal will be the original signal plus an additional undesired waveform whose spectrum overlaps with the high frequency components of the original signal. This undesired component is called aliasing noise and the overall effect is referred to as aliasing since the noise introduced here is actually a part of the signal itself but with its frequency components shifted to a new frequency.

The rate at which a signal is sampled usually determines the amount of processing, transmission or storage that will subsequently be required. Hence, it is desirable to use the lowest possible sampling rate that can satisfy a given application and does not violate the sampling theorem. On the other hand, the contribution of the higher frequency signal components usually diminishes in importance as frequency increases over certain values. For example, human eyes are not very sensitive to high frequency of color components Cb and Cr of an image [2-3]. Therefore, it is also important to choose a meaningful sampling rate that is not higher than necessary for the application.

The answer is to first decide how much of the original signal spectrum that really needs to be retained. Then an analog low-pass filtering is performed on the analog signal before sampling so that the "needless" high frequency components are suppressed. This analog prefiltering is often called antialias filtering. For example, in digital telephony, the standard antialias filter has a cutoff of 3.4 kHz although the speech signal contains frequency components extending well beyond this frequency. This cutoff allows the moderate sampling rate of 8 kHz to be used and retains the voice fidelity that was already achieved with analog telephone circuits that already were limited to roughly 3.4 kHz. In summary, analog prefiltering is needed to prevent aliasing of signal and noise components that lie outside of the frequency band that must be preserved and reproduced.

Just as a waveform is sampled at discrete times, the value of the sampled waveform at a given time is also converted to a discrete value. Such a conversion process is called quantization that will introduce loss on sampled waveform. The resolution of quantization depends on the number of bits used in measuring the height of the waveform. For example, an 8-bit quantization yields 256 possible values. The lower the resolution of quantization, the higher the loss of the digital signal. The electronic device that converts a signal waveform into digital samples is called the Analog-to-Digital (A/D) Converter. The reverse-conversion device is called a Digital-to-Analog (D/A) Converter.

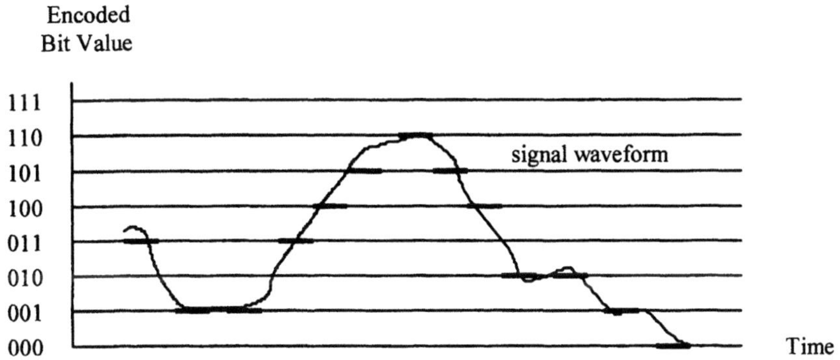

coded signal : 011 001 001 011 100 101 110 101 100 010 010 001 000

Figure 2.1 An Example of PCM Coded Signal of Sampled Waveform

The process which first samples analog signal and then quantizes the sample values is called pulse code modulation (PCM). Fig. 2.1 depicts an example of the steps involved in PCM at a high level. PCM does not require sophisticated signal processing techniques and related circuitry. Hence, it was the first method to be employed, and is the prevalent method used today in telephone plant. PCM provides excellent quality. The problem with PCM is that it requires a fairly high bandwidth to code a signal.

Two newer techniques, the differential pulse code modulation (DPCM) and adaptive DPCM (ADPCM), are among the most promising techniques for improving PCM at this time. If a signal has a high correlation between adjacent samples, the variance of the difference between adjacent samples is smaller than the variance of the original signal. If this difference is coded, rather than the original signal, fewer bits are needed for the same desired accuracy. That is, it is sufficient to represent only the first PCM-coded sample as a whole and all following samples as the difference from the previous one. This is the basic idea behind DPCM. In general, fewer bits are needed for DPCM than for PCM.

In a typical DPCM system, the input signal is band-limited, and an estimate of the previous sample (or a prediction of the current signal value) is subtracted from the input. The difference is then sampled and coded. In the simplest case, the estimate of the previous sample is formed by taking the sum of the decoded values of all the past differences (which ideally differ from the previous sample only by a quantizing error). DPCM exhibits a significant improvement over PCM when the signal spectrum is peaked at the lower frequencies and rolls off toward the higher frequencies.

A prominent adaptive coding technique is ADPCM. It is a successive development of DPCM. Here, differences are encoded by a use of a small number of bits only (e.g. 4 bits). Therefore, either sharp "transitions" are coded correctly (these bits represent bits with a higher significance) or small changes are coded exactly (DPCM-encoded values are the less-significant bits). In the second case, a loss of high frequencies would occur. ADPCM adapts to this "significance" for a particular data stream as follows: the coder divides the value of DPCM samples by a suitable coefficient and the decoder multiplies the compressed data by the same coefficient, i.e., the step size of the signal changes.

The value of the coefficient is adapted to the DPCM-encoded signal by the coder. In the case of a high-frequency signal, large DPCM coefficient values occur. The coder determines a high value for the coefficient. The result is a very coarse quantization of the DPCM signal in passages with steep edges. Low-frequency portions of such passages are hardly considered at all. For a signal with permanently relatively small DPCM values, the coder will determine a small coefficient. Thereby, a fine resolution of the dominant low frequency signal portions is guaranteed. If high-frequency portions of the signal suddenly occur in such a passage, a signal distortion in the form of a slope-overload arises. Considering the actually defined step size, the greatest possible change by a use of the existing number of bits will not be large enough to represent the DPCM value with an ADPCM value. The transition of the PCM signal will be faded.

It is possible to explicitly change the coefficient that is adaptively adjusted to the data in the coding process. Alternatively, the decoder is able to calculate the coefficients itself from an ADPCM-encoded data stream. In ADPCM, the coder can be made to adapt to DPCM value change by increasing or decreasing the range represented by the encoded bits. In principle, the range of bits can be increased or decreased to match different situations. In practice, the ADPCM coding device accepts the PCM coded signal and then applies a special algorithm to reduce the 8-bit samples to 4-bit words using only 15 quantization levels. These 4-bits words no longer represent sample amplitudes; instead, they contain only enough information to reconstruct the amplitude at the distant end. The adaptive predictor predicts the value of the next signal on the level of the previously sampled signal. A feedback loop ensures that signal variations are followed with minimal deviation. The deviation of the predicted value measured against the actual signal tends to be small and can be encoded with 4-bits.

2.3 Fundamental Compression Algorithms

The purpose of compression is to reduce the amount of data for multimedia communication. The amount of compression that an encoder achieves can be measured in two different ways. Sometimes the parameter of interest is compression ratio --- the ratio between the original source data and the compressed data sizes. However, for continuous-tone images another measure, the average number of compressed bits/pixel, is sometimes a more useful parameter for judging the performance of an encoding system. For a given image, however, the two are simply different ways of expressing the same compression.

Compression in multimedia systems is subject to certain constraints. The quality of the coded, and later on, decoded data should be as good as possible. To make a cost-effective implementation possible, the complexity of the technique should be minimal. The processing period of the algorithm cannot exceed certain time spans.

A natural measure of quality in a data coding and compression system is a quantitative measure of distortion. Among the quantitative measures, a class of criteria used often is called the mean square criterion. It refers to some sort of average or sum (or integral) of squares of the error between the sampled data $y(t)$ and decoded or decompressed data $y'(t)$. For data sequences $y(t)$ and $y'(t)$ of N samples, the quantity

$$\sigma_{ls}^2 = \frac{1}{N}\sum_{t=1}^{N}\left|y(t) - y'(t)\right|^2 \tag{2.3}$$

is called the *average least squares error*(ALSE). The quantity

$$\sigma_{ms}^2 = E\left[\left|y(t) - y'(t)\right|^2\right] \tag{2.4}$$

is called the mean square error (MSE), where E represents the mathematical expectation. Often ALSE is used as an estimate of MSE. In many applications the (mean square) error is expressed in terms of a signal-to-noise ratio (SNR), which is defined in decibels (dB) as

$$SNR = 10\log_{10}\frac{\sigma^2}{\sigma_e^2}, \quad \sigma_e = \sigma_{ms}, \sigma_{ls} \tag{2.5}$$

where σ^2 is the variance of the original sampled data sequence.

Another definition of SNR, used commonly in image and video coding applications, is

$$PSNR = 10\log_{10}\frac{(peak-to-peak\ value\ of\ the\ reference\ image)^2}{\sigma_e^2}. \tag{2.6}$$

The PSNR value is roughly 12 to 15 dB above the value of SNR.

Another commonly used method for performance measure of data coding and compression system is so-called rate distortion theory. Rate distortion theory provides some useful results, which tell us the minimum number of bits required to encode the data, while admitting a certain level of distortion and vice versa.

The rate distortion function of a random variable x gives the minimum average rate R_D (in bits per sample) required to represent (or code) it while allowing a fixed distortion D in its reproduced value. If x is a Gaussian random variable of variance σ^2 and y is its reproduced value and if the distortion is measured by the mean square value of the difference *(x-y)*, i.e., $D = E\left[(x-y)^2\right]$, then rate distortion function of x is defined as

$$R_D = \max\left\{0, \left(\frac{1}{2}\right)\log_2\left(\frac{\sigma^2}{D}\right)\right\}. \tag{2.7}$$

Data coding and compression systems are considered optimal if they maximize the amount of compression subject to an average or maximum distortion.

The quality of decompressed digital video is measured by three elements. These elements are the number of displayable colors, the number of pixels per frame (resolution), and the number of frames per second. Each of these elements can be traded off for another and all of them can be traded for better transmission rates.

Entropy Coding	Source Coding		Hybrid Coding
Huffman Coding	Predictions	PCM DPCM ADPCM	JPEG
Arithmetic Coding	Transformation	DCT FFT	MPEG-1 MPEG-2 MPEG4
Run-length Coding	Layered Coding		H.261 H.263
		Vector Quantization	

Table 2.1. A Classification of Coding/Compression Techniques

As shown in Table 2.1, compression techniques fit into different categories. For their use in multimedia systems, we can distinguish among *entropy, source,* and *hybrid coding*. Entropy coding is a lossless process, while source encoding is a lossy process. Most multimedia systems use hybrid techniques, which are a combination of the two.

Entropy coding is used regardless of the media data specific characteristics. Any input data sequence is considered to be a simple digital sequence and the semantics of the data is ignored. Entropy encoding reduces the size of the data sequence by focusing on the statistical characteristics of the encoded data to allocate efficient codes, independent of the characteristics of the data. Entropy encoding is an example of lossless encoding as the decompression process regenerates the data completely.

The basic ideas of entropy coding are as follows. First, we define the term *information* by using video signals as examples. Consider a video sequence in which each pixel takes on one of K values. If the spatial correlation have been removed from the video signal, the probability that a particular level i

appears will be P_i, independent of the spatial position. When such a video signal is transmitted, the information I imparted to the receiver by knowing which of K levels is the value of a particular pixel, is $-\log_2 P_i$ bits. This value, averaged over an image, is referred to as the average information of the image, or the *entropy*. The entropy can therefore be expressed as

$$H = -\sum_{i=0}^{K-1} P_i \log_2 P_i \ . \tag{2.8}$$

The entropy is also extremely useful for measuring the performance of a coding system. In "stationary" systems -- systems where the probabilities are fixed -- it provides a fundamental lower bound, what is called the entropy limit, for the compression that can be achieved with an alphabet symbol.

Entropy encoding attempts to perform efficient code allocation (without increasing the entropy) for a signal. Run-length encoding, Huffman encoding and arithmetic encoding are well-known entropy coding methods [2-7] for efficient code allocation, and are commonly used in actual encoders.

Run-length coding is the simplest entropy coding. Data streams often contain sequences of the same bytes or symbols. By replacing these repeated byte or symbol sequences with the number of occurrences, a substantial reduction of data can be achieved. This is called run-length coding, which is indicated by a special flag that does not occur in the data stream itself. For example, the data sequence: GISSSSSSSGIXXXXXX can be run-length coded as: GIS#7GIX#6 where # is the indicator flag. The character "S" occurs 7 consecutive times and is "compressed" to 3 characters "S#7" as well as the character "X" occurs 6 consecutive times and is also "compressed" to 3 characters "X#6". Run-length coding is a generalization of zero suppression, which assumes that just one symbol appears particularly often in sequences and the coding focuses on uninterrupted sequences, or *runs,* of zeros or ones to produce an efficient encoding.

Huffman coding is an optimal way of coding with integer-length code words. The Huffman coding produces a "compact" code. For a particular set of symbols and probabilities, no other integer code can be found that will give better coding performance than this compact code. Consider the example given in Table 2.2. The entropy -- the average ideal code length required to transmit the weather -- is given by

$$H = (1/16) \times 4 + (1/16) \times 4 + (1/8) \times 3 + (3/4) \times 0.415 = 1.186$$

bits/symbol.

Symbols	Probability(P)	Information (Ideal Length $-\log_2 P$)	Integer Code Length	Huffman Code
00	3/4	0.415 bits	1 bit	0
01	1/8	3 bits	2 bits	10
10	1/16	4 bits	3 bits	110
11	1/16	4-bits	3 bits	111

Table 2.2 Sequence Symbols with (1) Ideal Code Length, (2) Integer Code Length, and (3) Variable-length (Huffman) Codes

However, fractional-bit lengths are not allowed, so the lengths of the codes listed in the column to the right do not match the ideal information. Since an integer code always needs at least one bit, increasing the code for the symbol "00" to one bit seems logical.

The Huffman code assignment procedure is based on a coding "tree" structure. This tree is developed by a sequence of parsering operations in which the two least probable symbols are joined at a "node" to form two "branches" of the tree. As the tree is constructed, each node at which two branches meet is treated as a single symbol with a combined probability that is the sum of the probabilities for all symbols combined at that node.

Fig. 2.2 shows a Huffman code pairing sequence for the four-symbol case in Table 2.2. In this figure the four symbols are placed on the number line from 0 to 1 in order of increasing probability. The cumulative sum of the symbol probabilities is shown at the left. The two smallest probability intervals are paired, leaving three probability intervals of size 1/8, 1/8, and 3/4. We establish the next branch in the tree by again pairing the two smallest probability intervals, 1/8 and 1/8, leaving two probability intervals, 1/4 and 3/4. Finally, we complete the tree by pairing the 1/4 and 3/4 intervals. To create the code word for each symbol, we assign a 0 and 1, respectively (the order is arbitrary), to each branch of the tree. We then concatenate the bits assigned to these branches, starting at the "root" (at the right of the tree) and the following the branches back to the "leaf" for each symbol (at the far left). Notice that each node in this tree requires a binary decision -- a choice between the two possibilities -- and therefore appends one bit to the code word.

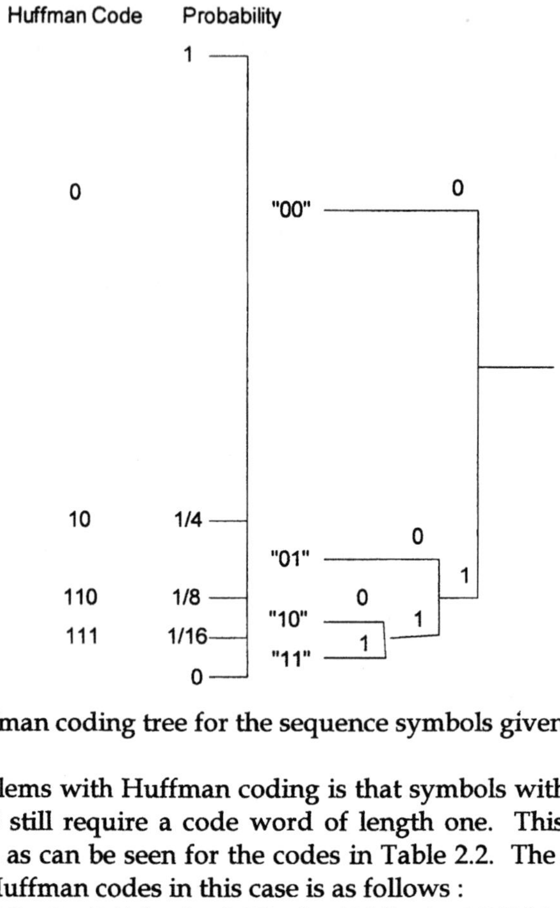

Figure 2.2 Huffman coding tree for the sequence symbols given in Table 2.2.

One of the problems with Huffman coding is that symbols with probabilities greater than 0.5 still require a code word of length one. This leads to less efficient coding, as can be seen for the codes in Table 2.2. The coding rate R achieved with Huffman codes in this case is as follows :

$$R = (1/16)\times 3 + (1/16)\times 3 + (1/8)\times 2 + (3/4)\times 1 = 1.375 \text{ bits/pixel.}$$

This rate, when compared to the entropy limit of 1.186 bits/pixel, represents an efficiency of 86%.

Arithmetic coding is an optimal coding procedure that is not constrained to integer-length codes. In arithmetic coding the symbols are ordered on the number line in the probability interval from 0 to 1 in a sequence that is known to both encoder and decoder. Each symbol is assigned a subinterval equal to its probability. Note that since the symbol probabilities sum to one, the subintervals precisely fill the symbol probabilities in Table 2.2. Fig. 2.3 illustrates a possible ordering for the symbol probabilities in Table 2.2.

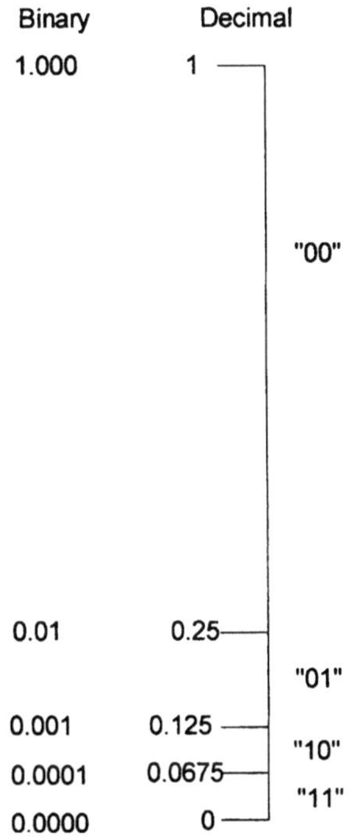

Figure 2.3 A process of partitioning the numbered line into subintervals for arithmetic coding.

The objective in arithmetic coding is to create a code stream that is a binary fraction pointing to the interval for the symbol being coded. Thus, if the symbol is "00", the code stream is a binary fraction greater than or equal to binary 0.01 (decimal 0.25), but less than binary 1.0. If the symbol is "01", the code stream is greater than or equal to binary 0.001, but less than binary 0.01. If the symbol is "10", the code stream is greater than or equal to binary 0.0001, but less than binary 0.001. Finally, if the symbol is "11", the code stream is greater than or equal to binary 0, but less than 0.0001. If the code stream follows these rules, a decoder can see which subinterval is pointed to by the code stream and decode the appropriate symbol. Coding additional symbols is a matter of subdividing the probability interval into smaller and smaller subintervals, always in proportion to the probability of the particular symbol sequence. As long as we follow the rules never allow the code stream to point outside the subinterval assigned to the sequence of symbols, the decoder will decode that sequence.

For a detailed discussion of Huffman coding and arithmetic coding, interested readers should refer to reference [2-7].

Source coding takes into account the semantics of the data. The degree of compression that can be reached by source coding depends on the data contents. In the case of lossy compression techniques, a one-way relation between the original sequence and the encoded data stream exists; the data streams are similar but not identical. Different source coding techniques make extensive use of the characteristics of the specific medium. An example is the speech source coding, where speech is transformed from time-dependent to frequency-dependent speech concatenations, followed by the encoding. This transformation substantially reduces the amount of data.

Predictive Coding is the most fundamental source coding. The basis of predictive encoding is to reduce the number of bits used to represent information by taking advantage of correlation in the input signal. DPCM and ADPCM discussed above are among the simplest prediction coding methods. For digital video, signals exhibit correlation both between pixels within a picture (spatial correlation) and between pixels in differing pictures (temporal correlation). Video compression falls into two main types: (1) inter-picture prediction which uses combination of key motion-predicted and interpolated pictures to achieve high-compression ratio; (2) intra-picture coding which compress every picture of video individually. Inter-picture prediction techniques take advantage of the temporal correlation, while the spatial correlations are exploited by intra-picture coding methods. For interlaced video, it is amenable also to intra and inter-field picture prediction methods because interlaced video scans alternate lines to distribute the pixels of a single picture across two fields.

Motion compensation (MC), one of the most complex prediction methods, reduces the prediction error by predicting the motion of the imaged objects. The basic idea of MC arises from a common sense observation: in a video sequence, successive pictures are likely to represent the same details, with little difference between one picture and the next. A sequence showing moving objects over a still background is a good example. Data compression can be effected if each component of a picture is represented by its difference with the most similar component - the predictor - in the previous picture, and by a vector - the motion vector - expressing the relative position of the two components. If an actual motion exists between the two pictures, the difference may be null or very small. The original component can be reconstructed from the difference, the motion vector, and the previous picture.

Motion-compensated prediction is a powerful tool to reduce temporal redundancies between pictures and is used extensively in MPEG-1, MPEG-2 and MPEG-4 standards as the inter-picture coding technique. If all elements in a video scene are approximately spatially displaced, the motion between pictures can be represented by a number of motion parameters, e.g. by motion vectors for translation motion of pixels. Thus, the prediction of an actual pixel can be given by a motion-compensated prediction pixel from a previously coded picture. Usually both, prediction error and motion vectors, are transmitted to the receiver. However, encoding every motion vector with each coded picture pixel is generally neither desirable nor necessary. Since the spatial correlation between motion vectors is often high it is sometimes assumed that one motion vector is representative for the motion of a "block" of adjacent pixels. To this aim pictures are usually separated into disjoint blocks of pixels, e.g. 8x8 pixels in MPEG-4 and 16x16 pixels in MPEG-1, MPEG-2 and MPEG-4 standards, and only one motion vector is estimated, coded and transmitted for each of these blocks.

In the MPEG compression algorithms, the motion compensated prediction techniques are used for reducing temporal redundancies between pictures and only the prediction error pictures - the difference between original pictures and motion compensated prediction pictures - are encoded. In general the correlation between pixels in the motion-compensated inter-picture error pictures to be coded is reduced compared to the correlation properties of intra-pictures due to the prediction based on the previous coded picture.

A weakness of prediction-based encoding is that the influence of any errors during data transmission affects all subsequent data. In particular, when inter-picture prediction is used, the influence of transmission errors is quite noticeable. Since predictive encoding schemes are often used in combination with other schemes, such as transform-based schemes, the influence of transmission errors must be given due consideration.

Transform Coding has been studied extensively during the last two decades and has become a very popular compression method for still picture coding and video coding. The purpose of transform coding is to de-correlate the intra- or inter-picture error picture content and to encode transform coefficients rather than the original pixels of the pictures. To this aim the input pictures are split into disjoint blocks of pixels â (i.e. of size *N*x*N* pixels). The transformation can be represented as a matrix operation using a *N*x*N* transform matrix A to obtain the NxN transform coefficients c based on a linear, separable and unitary *forward* transformation

$$c = A\,\hat{a}\,A^T.$$

Here, A^T denotes the transpose of the transformation matrix A. Note, that the transformation is reversible, since the original NxN block of pixels â can be reconstructed using a linear and separable inverse transformation

$$\hat{a} = A^T\,c\,A.$$

A major objective of transform coding is to make many Transform coefficients small enough so that they are insignificant in terms of both statistical and subjective measures and need not be coded for transmission. At the same time it is desirable to minimize statistical dependencies between coefficients with the aim to reduce the amount of bits needed to encode the remaining coefficients.

Upon many possible alternatives the Discrete Cosine Transform (DCT) applied to smaller picture blocks of usually *8x8* pixels has become the most successful transform for still picture and video coding [2-8]. In fact, DCT based implementations are used in most picture and video coding standards due to their high de-correlation performance and the availability of fast DCT algorithms suitable for real time implementations. The standards that use 8x8 DCT are H.261, H.263, MPEG-1, MPEG-2, MPEG-4 part2, and JPEG. VLSI implementations that operate at rates suitable for a broad range of video applications are commercially available today.

The 1-dimensional DCT transform maps a length-N vector **x** into a new vector **X** of transform coefficients by a linear transformation **X** = **H x**, where the element in the kth row and nth column of **H** is defined by

$$\mathbf{H}_{kn} = H(k,n) = c_k \sqrt{\frac{2}{N}} \cos\left((n + 1/2) \frac{k\pi}{N} \right) \qquad (2.9)$$

for k = 0, 1, ..., N-1, and n = 0, 1, ..., N-1, with $c_0 = \sqrt{2}$ and $c_k = 1$ for $k > 1$. The DCT matrix is orthogonal, so its inverse equals its transpose, that is $\mathbf{x} = \mathbf{H}^{-1}\mathbf{X} = \mathbf{H}^T\mathbf{X}$.

The following expresses a 2-dimensional DCT for an N x N pixel block.

$$X(u,v) = \frac{4C(u)C(v)}{N^2} \sum_{j=0}^{N-1} \sum_{k=0}^{N-1} x(j,k) \cos\left(\frac{(2j+1)u\pi}{2N} \right) \cos\left(\frac{(2k+1)v\pi}{2N} \right)$$

$$x(j,k)=\sum_{u=0}^{N-1}\sum_{v=0}^{N-1}C(u)C(v)X(u,v)\cos\left(\frac{(2j+1)u\pi}{2N}\right)\cos\left(\frac{(2k+1)v\pi}{2N}\right) \tag{2.10}$$

where

$$C(w)=\begin{cases}\frac{1}{\sqrt{2}}, & w=0\\ 1, & w=1,2,\dots,N-1.\end{cases}$$

After the transformation, output coefficients are quantized by levels specified in a quantization table. Usually, larger values of N improve the SNR, but the effect saturates above a certain block size. Further, increasing the block size increases the total computation cost required. The value of N is thus chosen to balance the efficiency of the transform and its computation cost, block sizes of 4 and 8 are common. For large quatization, segmentation DCT into size 8 blocks often leads to "blocking artifacts" -- visible discontinuities between adjacent blocks. However, the blocking artifacts are less visible for the DCT transform of size 4.

The DCT is closely related to Discrete Fourier Transform (DFT) and it is of some importance to realize that the DCT coefficients can be given a frequency interpretation close to the DFT. Thus low DCT coefficients relate to low spatial frequencies within picture blocks and high DCT coefficients to higher frequencies. This property is used in many coding schemes to remove subjective redundancies contained in the picture data based on human visual systems criteria. Since the human viewer is more sensitive to reconstruction errors related to low spatial frequencies than to high frequencies, a frequency adaptive weighting (quantization) of the coefficients according to the human visual perception (perceptual quantization) is often employed to improve the visual quality of the decoded pictures for a given bit rate.

Next, we will discuss an integer approximation of DCT. One disadvantage of the DCT is that the entries $H(k, n)$ in Eq.(2.9) are irrational numbers, and so integer input data $x(n)$ will map to irrational transform coefficients $X(k)$. Thus, in a digital computer, when we compute the direct and inverse transform in cascade, we do not get exactly the same data back. In other words, if we compute $\mathbf{X} = \mathbf{H}\,\mathbf{x}$ and $\mathbf{u}$ = rounding($\mathbf{H}^T \mathbf{X}$), then it is not true that $u(n) = x(n)$ for all n. If we introduce appropriate scale factors a, e.g. in $\mathbf{X} = a\,\mathbf{H}\,\mathbf{x}$ and $\mathbf{u}$ = rounding($a\,\mathbf{H}^T \mathbf{X}$), then we can make $u(n) = G\,x(n)$, where G is an integer, for almost all n by choosing a large enough and a appropriately. Nevertheless, an exact result cannot be guaranteed.

In a motion-compensated video encoder, past decoded frames are used as reference information for prediction of the current frame. Therefore, the encoder has to generate such decoded frames, and for that it needs to compute inverse transforms. If the formula **u** = rounding(a $\mathbf{H}^T$ **X**) is used, then different floating-point formats and rounding strategies in different processors will lead to different results. That will result in a drift between the decoded data at the decoder and encoder.

One solution to the data drift problem is to approximate the matrix **H** by a matrix containing only integers. If the rows of **H** are orthogonal and have the same norm, then it follows that **u** can be computed exactly in integer arithmetic for all integer **x**. In other words, when we compute the direct transform by **X** = **H x** and the inverse transform by **u** = $\mathbf{H}^T$**X**, then we will have **u** = G **x**, where G is an integer equal to the squared norm of any of the rows in **H**.

Integer approximations to the DCT can be generated by trial-and-error, by approximating a scaled DCT matrix a**H** by integers [2-4][2-12]. Such approximations should preserve the symmetries in the rows of **H**. Clearly, a simple way to generate integer approximations to the DCT is by using the general formula

$$Q(k,n) = \text{rounding}(a\,H(k,n)),$$

where a is a scaling parameter. Let's consider N = 4 (note that this is the transform size in MPEG-4 part 10), for which the DCT matrix is given by

$$\mathbf{H} = \frac{1}{2}\begin{bmatrix} 1 & 1 & 1 & 1 \\ c & s & -s & -c \\ 1 & -1 & -1 & 1 \\ s & -c & c & -s \end{bmatrix},$$

where $c = c_0 \cdot \cos\left(\frac{\pi}{8}\right)$ and $s = c_0 \cdot \sin\left(\frac{\pi}{8}\right)$, where $c_0 = \sqrt{2}$.

For example, if a = 26, the transform matrix is

$$Q_0 = round(26\mathrm{H}) = \begin{bmatrix} 13 & 13 & 13 & 13 \\ 17 & 7 & -7 & -17 \\ 13 & -13 & -13 & 13 \\ 7 & -17 & 17 & -7 \end{bmatrix}.$$

Note that the rows and columns of $\mathbf{Q}_0$ are orthogonal to each other (the inner product of any two columns is zero), and all have norm equal to 26. In fact, for $a < 100$ we can only get orthogonal matrices with equal-norm rows by setting $a = 2$ or $a = 26$. The solution for $a = 2$ is not useful, since it's a Hadamard matrix [2-11], which does not lead to nearly as good compression as the DCT. Large values for a are not attractive because of the increase in the word length require to compute the results of the direct transform $\mathbf{X} = \mathbf{Q}_0\, \mathbf{x}$. We define the inverse transform by $\mathbf{x}' = \mathbf{Q}_0\, \mathbf{X}$, so it can also be computed with integer arithmetic. From the definition above, it is easy to see that $\mathbf{x}' = 676\, \mathbf{x}$, i.e. the reconstructed data $\mathbf{x}'$ is equal to the original data $\mathbf{x}$ amplified by an integer gain of 676 (which is the norm of any of the rows in $\mathbf{Q}_0$).

If $a = 2.5$, the transform matrix is

$$Q_0 = round(2.5\mathrm{H}) = \begin{bmatrix} 1 & 1 & 1 & 1 \\ 2 & 1 & -1 & -2 \\ 1 & -1 & -1 & 1 \\ 1 & -2 & 2 & -1 \end{bmatrix}.$$

In practice, DCT is used in conjunction with other techniques, such as prediction and entropy coding. The **Motion Compensation Plus Discrete Cosine Transform (MC + DCT)** scheme, which we will repeatedly refer to, is a prime example of such a combination.

MC + DCT*:* Suppose that the video to be encoded consists of digital television or teleconferencing services. For this type of video, MC carried out on the basis of picture differences is quite effective. MC can be combined with the DCT for even more effective compression. The overall configuration of MC + DCT is illustrated in Fig. 2.4. The selection of block size compares its input signal with that of the previous picture (generally in units of 8×8 pixel blocks) and selects those that exhibit motion. MC operates by comparing, the input signal in units of blocks against a locally decoded copy of the previous picture, extracting a motion vector and using the motion vector to calculate the picture difference. The motion vector is extracted by, for example, shifting vertically or horizontally a region several pixels on a side and performing matching within the block or the macroblock (a 16×16 pixel segment in a picture) [2-8].

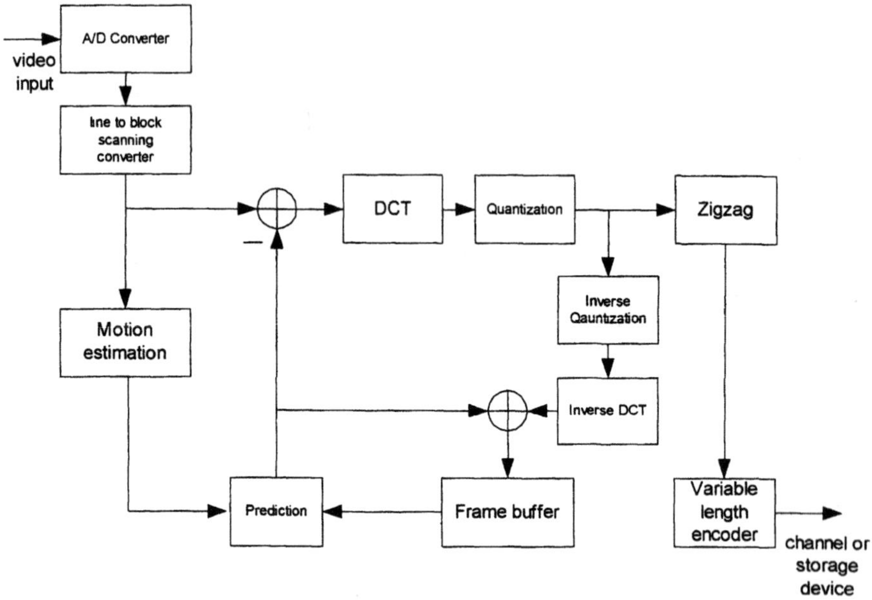

Figure 2.4 A block diagram of the MC+DCT coding scheme

The motion-compensated picture-difference signal is then transformed in order to remove spatial redundancy. A variety of compression techniques are applied in quantizing the transform coefficients; the reader is directed to the references for details [2-8]. A commonly-used method is zig-zag scan, which has been standardized in JPEG, H.261, H.263, MPEG-1, -2, and -4, for video transmission encoding [2-8]. Zig-zag scan, which transforms 2-dimensional data into one dimension, is illustrated in Fig. 2.5. Because the DC component of the coefficients is of critical importance, ordinary linear quantization is employed for them. Other components are scanned, for example in zig-zag fashion, from low to high frequency, linearly quantized, and variable-length-encoded by the use of run-length and Huffman coding.

Subband coding [2-5] refers to the compression methods that divide the signal into multiple frequency bands to take advantage of a bias in the frequency spectrum of the video signal. Efficient encoding is performed by partitioning the signal into multiple bands and taking into account the statistical characteristics and visual significance of each band.

The general form of a subband coding system is shown in Fig. 2.6. In the encoder, the analyzing filters partition the input signal into bands, each band is separately encoded, and the encoded bands are multiplexed and transmitted. The decoder reverses this process. Subband encoding does offer several advantages. Unlike DCT, it is not prone to blocking artifacts. Furthermore, subband encoding is the most natural coding scheme when hierarchical processing is needed for video coding.

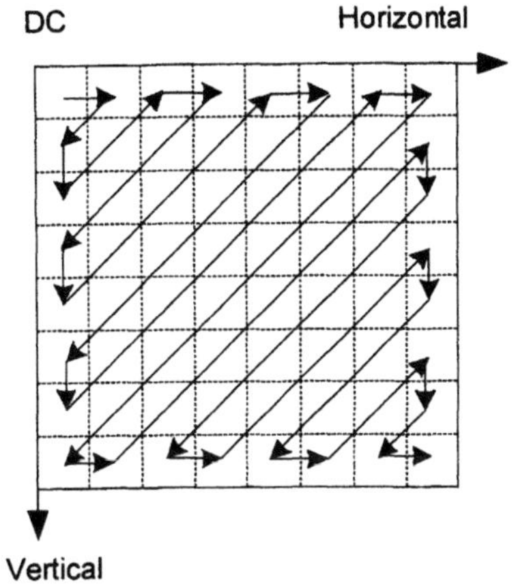

v \ u	0	1	2	3	4	5	6	7
0	0	1	5	6	14	15	27	28
1	2	4	7	13	16	26	29	42
2	3	8	12	17	25	30	41	43
3	9	11	18	24	31	40	44	53
4	10	19	23	32	39	45	52	54
5	20	22	33	38	46	51	55	60
6	21	34	37	47	50	56	59	61
7	35	36	48	49	57	58	62	63

Figure 2.5 The zigzag scan pattern for 8x8 block.

The main technological features to be determined in subband encoding are the subband analysis method (2- or 3-dimensional), the structure of the analyzing filters, the bit allocation method, and the compression method within each band. In particular, there are quite a number of candidates for the form of the analysis and the structure of the filters. The filters must not introduce distortion due to aliasing in-band analysis and synthesis.

Fig. 2.7 shows a 2-band analysis and synthesis system. Consider the following analyzing filter as an example:

$$H_l(z) = \frac{1}{4}\left(1 + 2z^{-1} + z^{-2}\right),$$

$$H_h(z) = \frac{1}{4}\left(-1 - 2z^{-1} + 6z^{-2} - 2z^{-3} - z^{-4}\right).$$

For these analyzing filters, the characteristics of the synthesizing filters are

$$G_l(z) = H_h(-z) \qquad G_h(z) = H_l(-z).$$

The relationship between the input and output is then

$$Y(z) = \frac{1}{2}\left[H_l(z)X(z) + H_l(-z)X(-z)\right]G_l(z)$$
$$-\frac{1}{2}\left[H_h(z)X(z) + H_h(-z)X(-z)\right]G_h(z)$$
$$= z^{-3}X(z)$$

Clearly, the aliasing components completely cancel. The basic principles illustrated hold unchanged when 2-dimensional filtering is used in a practical application.

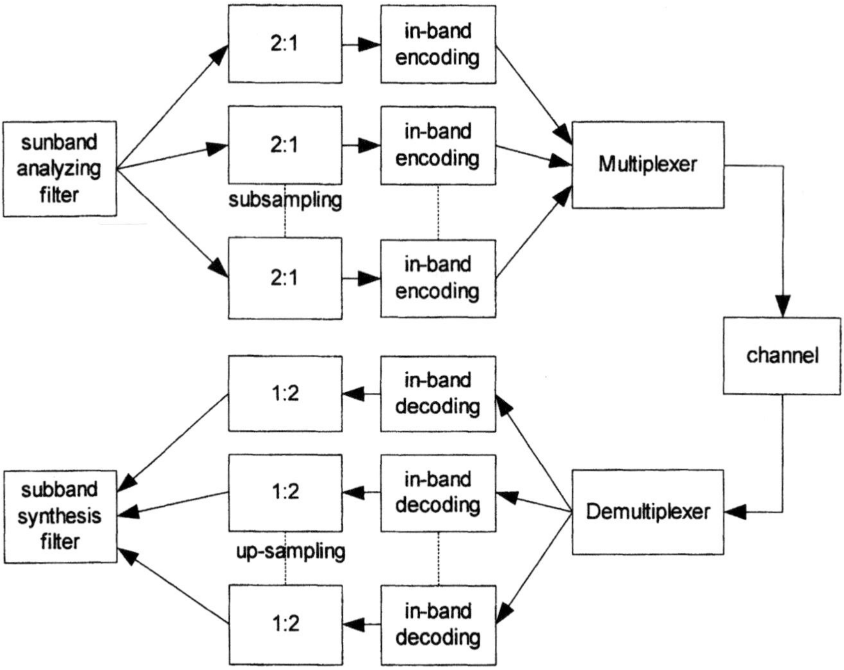

Figure 2.6 A simplified block diagram of subband coding scheme

Fig. 2.8 illustrates how the 2-dimensional frequency domain may be partitioned either uniformly or in an octave parent. If we recall that signal power will be concentrated in the low-frequency components, then the octave method seems the most natural. Since this corresponds to constructing the analyzing filters in a tree structure, it lends itself well to implementation with filter banks.

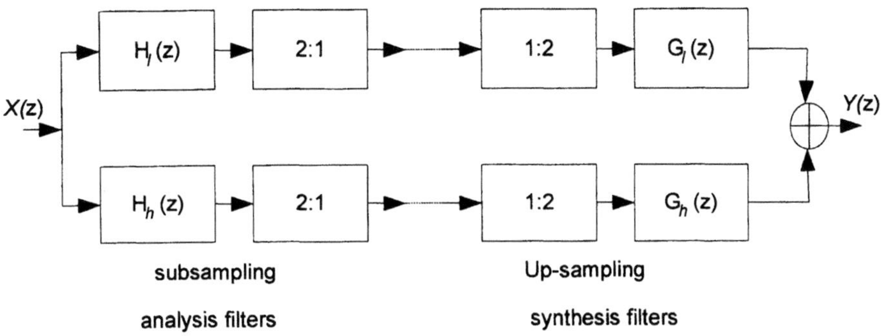

Figure 2.7 A two-band subband coding system. It demonstrates the in-band encoding and the in-band decoding blocks in Figure 2.6.

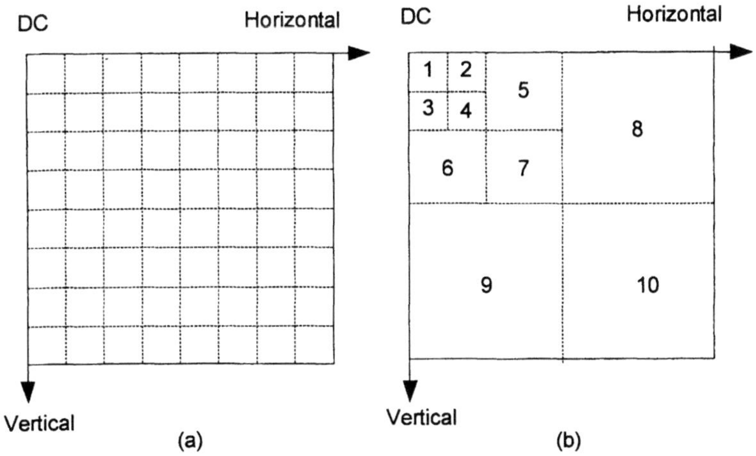

Figure 2.8 Subband splitting patterns in two-dimensional frequency domain: (a) uniform split (8x8); (b) octave split

The organization of a subband codec is similar to the DCT-based codec. The principal difference is that encoding and decoding are each broken out into a number of independent bands. Quality can be fixed at any desired value by adjusting the compression and quantization parameters of the encoders for each band. Entropy coding and predictive coding are often used in conjunction with subband coding to achieve high compression performance.

If we consider quality from the point of view of the rate-distortion curve then, at any given bit rate, the quality can be maximized by distributing the bits such that distortion is constant for all bands. A fixed number of bits is allocated, in advance, to each band's quantizer based on the statistical characteristics of the band's signal. In contrast, adaptive bit distribution adjusts the bit count of each band according to the power of the signal. In this case, either the decoder of each subband must also determine the bit count for inverse quantization, using the same criterion as is used by the encoder, or the bit count information must be transmitted along with the quantized signal. Therefore, the method is somewhat lacking in robustness.

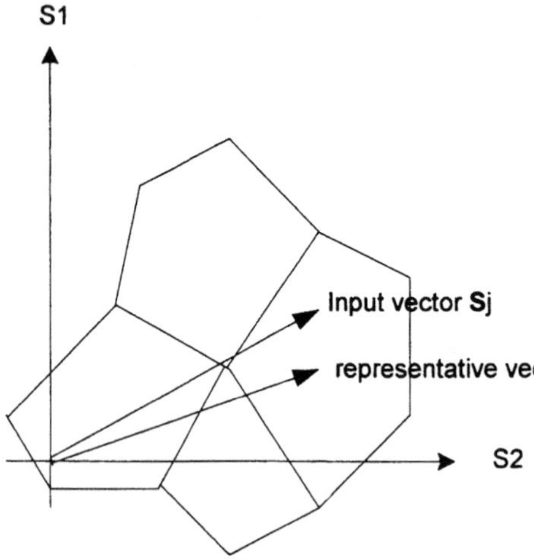

Figure 2.9 An example of VQ with two-dimensional vectors.

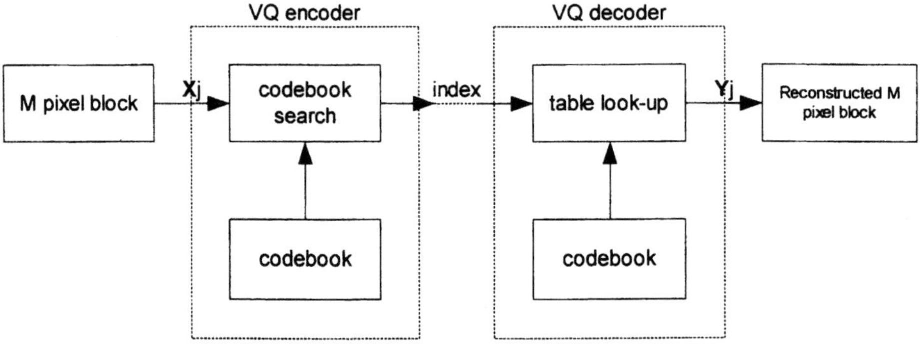

Figure 2.10 The basic structure of a VQ codec

Vector Quantization*:* As opposed to scalar quantization, in which sample values are independently quantized one at a time, vector quantization (VQ) attempts to remove redundancy between sample values by collecting several sample values and quantizing them as a single vector. Since the input to a scalar quantizer consists of individual sample values, the signal space is a finite interval of the real number line. This interval is divided into several regions, and each region is represented in the quantized outputs by a single value. The input to a vector quantizer is typically an n-dimensional vector, and the signal space is likewise an n-dimensional space. To simplify the discussion, we consider only the case where $n = 2$. In this case, the input to the quantizer is the vector S_j, which corresponds to the pair of samples (S_{1j}, S_{2j}). To perform vector quantization, the signal space is divided into a finite number of nonoverlapping regions, and a single vector to represent each region is determined. When the vector S_j is input, the region containing S_j is determined, and the representative vector for that region, Y_j, is output. This concept is shown in Fig. 2.9. If we phrase the explanation explicitly in terms of encoding and decoding, the encoder determines the region to which the input S_j belongs and outputs j, the index value that represents the region. The decoder receives this value j, extracts the corresponding vector Y_j from the representative vector set, and outputs it. The set of representative vectors is called the codebook.

The performance of vector quantization is evaluated in the same manner as for other schemes, that is, by the relationship between the encoding rate and the distortion. The encoding rate R per sample is given by the following equation,

$$R = \lceil \log_2 N \rceil / K, \tag{2.11}$$

where K is the vector dimensionality, and N is the number of quantization revels. The notation $\lceil \cdot \rceil$ represents the smallest integer greater than or equal to x (the "ceiling" of x).

We define the distortion as the distance between the input vector S_j and the output vector Y_j. In video encoding, the square of the Euclidean distance is generally used as a distortion measure because it makes analytic design of the vector quantizer for minimal distortion more tractable. However, it is not necessarily the case that subjective distortion perceived by a human observer coincides with the squared distortion.

To design a high performance vector quantizer, the representative vectors and the regions they cover must be chosen to minimize total distortion. If the input vector probability density function is known in advance, and the vector dimensionality is low, it is possible to perform an exact optimization. However, in an actual application it is rare for the input vector probability density to be known in advance. The well-known LBG algorithm is widely used for adaptively designing vector quantizers in this situation [2-9]. LBG is a practical algorithm that starts out with some reasonable codebook, and, by adaptively iterating the determination of regions and representative vectors, converges on a better codebook.

Fig. 2.10 shows the basic structure of an image codec based on vector quantization. The image is partitioned into M-pixel blocks, which are presented, one at a time, to the VQ encoder as the l-dimensional vector S_j. The encoder locates the closest representative vector in its prepared codebook and transmits the representative vector's index. The decoder, which need only perform a simple table lookup in the codebook to output the representative vector, is an extremely simple device. The simplicity of the decoder makes VQ coding very attractive for distribution-type video services. VQ coding, combining with other coding methods, has been adopted in many high-performance compression systems.

Table 2-1 shows examples of coding and compression techniques that are applicable in multimedia applications in relation to the entropy, source and hybrid coding classification. Hybrid compression techniques are a combination of well-known algorithms and transformation techniques that can be applied to multimedia systems. For a better and clearer understanding of hybrid schemes we will identify in all schemes (entropy, source and hybrid) a set of typical processing steps.

This typical sequence of operations has been shown in Fig. 2.4, which is performed in the compression of still images and video sequences. The following four steps describe single image compression.

1. Preparation includes analog-to-digital conversion and generating an appropriate digital representation of the information. For example, an image is divided into blocks of 8x8 pixels, and represented by a fixed number of bits per pixel.
2. Processing is actually the first step of the compression process that makes use of sophisticated algorithms. For example, a transformation from the time to the frequency domain can be performed by a use of DCT. In the case of motion video compression, inter-picture coding uses a motion vector for each 16x16 macroblock or 8x8 block.
3. Quantization processes the results of the previous step. It specifies the granularity of the mapping of real numbers into integers. This process results in a reduction of precision. In a transformed domain, the coefficients are distinguished according to their significance. For example, they could be quantized using a different number of bits per coefficient.
4. Entropy encoding is usually the last step. It compresses a sequential digital data stream without loss. For example, a sequence of zeros in a

data stream can be compressed by specifying the number of occurrences followed by the zero itself.

In the case of vector quantization, a data stream is divided into blocks of *n* bytes each. A predefined table contains a set of patterns. For each block, a table entry with the most similar pattern is identified. Each pattern in the table is associated with an index. Such a table can be multi-dimensional; in this case, the index will be a vector. A decoder uses the same table to generate an approximation of the original data stream.

2.4 Image and Video Compression Standards

In the following sections the most relevant work in the standardization bodies concerning image and video coding is outlined. In the framework of International Standard Organization (ISO/IEC/JTC1), four subgroups were established in May 1988: JPEG (Joint Photographic Experts Group) is working on coding algorithms for still images; JBIG (Joint Bi-level Expert Group) is working on the progressive processing of bi-level coding algorithms, and MPEG (Moving Picture Experts Group) is working on representation of motion video. In the International Telecommunication Union (ITU), H.261 and H.263 are also developed for video conferencing and telephone applications. The results of these standard activities are presented next.

JPEG: The ISO 10918-1 JPEG International Standard (1992) | CCITT (former ITU) Recommendation T.81 is a standardization of compression and decompression of still natural images [2-4]. JPEG provides the following important features:

- JPEG implementation is independent of image size.
- JPEG implementation is applicable to any image and pixel aspect ratio.
- Color representation is independent of the special implementation.
- JPEG is for natural images, but image content can be of any complexity, with any statistical characteristics.
- The encoding and decoding complexities of JPEG are balanced and can be implemented by a software solution.
- Sequential decoding (slice-by-slice) and progressive decoding(refinement of the whole image) should be possible. A lossless, hierarchical coding of the same image with different resolutions is supported.

- The user can select the quality of the reproduced image, the compression processing time and the size of the compressed image by choosing appropriate individual parameters.

The key steps of the JPEG compression are DCT (8×8), quantization, zig-zag scan, and entropy coding. Both Huffman coding and arithmetic coding are options of entropy coding in JPEG. The JPEG decompression just reverses its compression process. A fast coding and decoding of still images is also used for video sequences known as Motion JPEG. Today, JPEG software packages or together with specific hardware support are already available in many products.

ISO 11544 JBIG is specified for lossless compression of binary and limited bits/pixel images [2-4]. The basic structure of the JBIG compression system is an adaptive binary arithmetic coder. The arithmetic coder defined for JBIG is identical to the arithmetic-coder option in JPEG.

Most recently, JPEG has developed a new wavelet-based codec, namely JPEG-2000. Such a codec can provide much higher coding performance. However, the complexity of the codec is also very high.

H.261 and H.263: ITU Recommendations H.261 and H.263 [2-6] are digital video compression standards that are developed for video conferencing and videophone applications, respectively.

Both H.261 and H.263 are developed for real-time encoding and decoding. For example, the maximum signal delay of both compression and decompression for H.261 is specified as 150 milliseconds by the end-to-end delay of targeted applications. Unlike JPEG, H.261 and H.261 specify a very precise image format. Two resolution formats each with an aspect ratio of 4:3 are specified. the so-called Common Intermediate Format (CIF) defines a luminance component (Y) of 288 lines, each with 352 pixels. The chrominance components (C_b and C_r) have a solution with a rate of 144 lines and 176 pixels per line to fulfill the 2:1:1 requirement. Quarter-CIF (QCIF) has exactly half of the CIF resolution, i.e., 176×144 pixels for the luminance and 88×72 pixels for the other components. All H.261 implementation must be able to encode and decode QCIF.

In H.261 and H.263, data units of the size 8×8 pixels are used for the representation of the Y, as well as the C_b and C_r components. A macroblock is the result of combining four Y blocks with one block of the C_b

and C_r components. A group of blocks is defined to consist of 33 macroblocks. Therefore, a QCIF-image consists of three groups of blocks, and a CIF-image comprises twelve groups of blocks. Two types of pictures are considered in the H.261 coding. These are I-pictures (or intraframes) and P-pictures (or interframes). For I-picture encoding, each macroblock is intra-coded. That is, each block of 8×8 pixels in a macroblock is transformed into 64 coefficients by a use of DCT and then quantized. The quantization of DC-coefficients differs from that of AC-coefficients. The next step is to apply entropy encoding to the DC- and AC-parameters, resulting in a variable-length encoded word. For P-picture encoding, the macroblocks are either MC+DCT coded or intra-coded. The prediction of MC+DCT coded macroblocks is determined by a comparison of macroblocks from previous images and the current image. Subsequently, the components of the motion vector are entropy encoded by a use of a lossless variable-length coding system. To improve the coding efficiency for low bit-rate applications, several new coding tools are included in H.263. Among them are the PB-picture type and overlapped motion compensation, etc..

The combination of the temporal motion-compensated prediction and transform domain coding can be seen as the key elements of the MPEG coding standards. To this reason the MPEG coding algorithms are usually referred to as hybrid block-based DPCM/DCT algorithms.

MPEG-1 [2-16] is a generic standard for coding of moving pictures and associated audio for digital storage media at up to about 1.5 Mbits/s. The video compression technique developed in MPEG-1 covers many applications from interactive VCD to the delivery of video over telecommunications networks. The MPEG-1 video coding standard is thought to be generic. To support the wide range of applications profiles a diversity of input parameters including flexible picture size and rate that can be specified by the user. MPEG has recommended a constraint parameter set: every MPEG-1 compatible decoder must be able to support at least video source parameters up to TV size: including a minimum horizontal size of 720 pixels, a minimum vertical size of 576 pixels, a minimum picture rate of 30 pictures per second and a minimum bit rate of 1.86 Mbits/s. The standard video input consists of a non-interlaced video picture format. But, it should be noted that by no means the application of MPEG-1 is limited to this constrained parameter set.

The MPEG-1 video algorithm has been developed with respect to the JPEG [2-5] and H.261 [2-6] activities. It was intended to retain a large degree of commonalty with the H.261 standard so that implementations supporting both standards were plausible. However, MPEG-1 was primarily targeted for multimedia CD-ROM applications, requiring additional functionality supported by both encoder and decoder. Important features provided by MPEG-1 include picture based random access of video, fast forward/fast reverse *(FF/FR)* searches through compressed bit streams, reverse playback of video and editing ability of the compressed bit stream.

The Basic MPEG-1 Inter-Picture Coding Scheme. The basic MPEG-1 (as well as the MPEG-2) video compression technique is based on a Macroblock structure, motion compensation and the conditional replenishment of Macroblocks. As outlined in Fig. 2.11a the MPEG-1 coding algorithm encodes the first picture in a video sequence in Intra-picture coding mode (I-picture). Each subsequent picture is coded using Inter-picture prediction (P-pictures) - only data from the nearest previously coded I- or P-picture is used for prediction. The MPEG-1 algorithm processes the pictures of a video sequence block-based. Each colour input picture in a video sequence is partitioned into non-overlapping "Macroblocks" as depicted in Fig. 2.11b. Each Macroblock contains blocks of data from both luminance and co-sited chrominance bands - four luminanceblocks (Y1, Y2, Y3, Y4) and two chrominance blocks (U, V), each with size *8 x 8* pels. Thus the sampling ratio between Y:U:V luminance and chrominance pixels is 4:1:1.

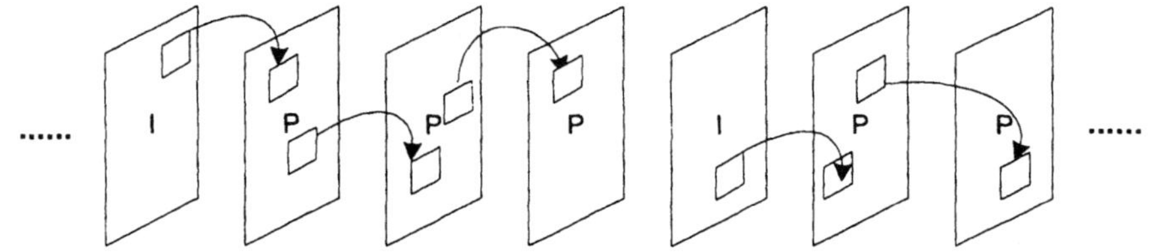

Figure 2.11a An example of MPEG-1 coding streucture

P-pictures are coded using motion compensated prediction based on the nearest previous picture. Each picture is divided into disjoint "Macroblocks" (MB). With each Macroblock (MB), information related to four luminance blocks (Y1, Y2, Y3, Y4) and two chrominance blocks (U, V) is coded. Each block contains 8x8 pels.

The block diagram of the basic hybrid DPCM/DCT MPEG-1 encoder and decoder structure is depicted in Fig. 2.5. The first picture in a video sequence (I-picture) is encoded in INTRA mode without reference to any past or future pictures. At the encoder the DCT is applied to each *8 x 8* luminance and chrominance block and, after output of the DCT, each of the 64 DCT coefficients is uniformly quantized (Q) . The quantizer stepsize (sz) used to quantize the DCT-coefficients within a Macroblock is transmitted to the receiver. After quantization, the lowest DCT coefficient (DC coefficient) is treated differently from the remaining coefficients (AC coefficients). The DC coefficient corresponds to the average intensity of the component block and is encoded using a differential DC prediction method. The non-zero quantizer values of the remaining DCT coefficients and their locations are then "zig-zag" scanned and run-length entropy coded using variable length code (VLC) tables.

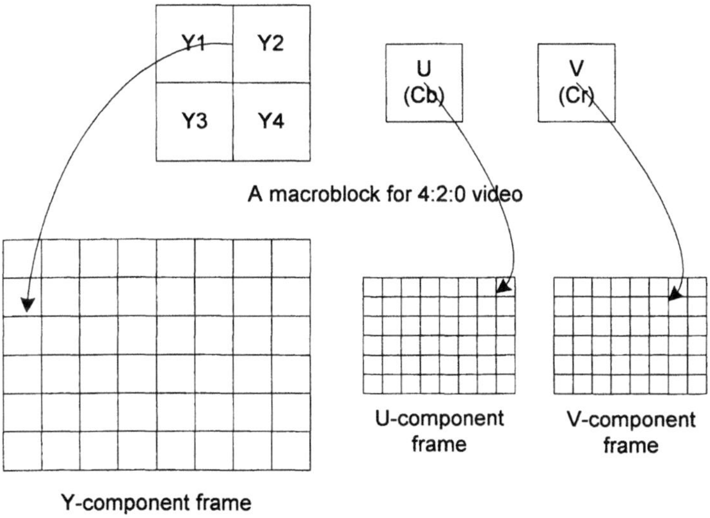

Figure 2.11b Macroblock structure and YUV (or Y:Cb:Cr) frame

The concept of "zig-zag" scanning of the coefficients is outlined in Fig. 2.6. The scanning of the quantized DCT-domain 2-dimensional signal followed by variable-length code-word assignment for the coefficients serves as a mapping of the 2-dimensional picture signal into a 1-dimensional bitstream. The non-zero AC coefficient quantizer values (length,) are detected along the scan line as well as the distance (run) between two consecutive non-zero coefficients. Each consecutive (run, length) pair is encoded by transmitting only one VLC codeword. The purpose of "zig-zag" scanning is to trace the low-frequency DCT coefficients (containing most energy) before tracing the high-frequency coefficients.

The decoder performs the reverse operations, first extracting and decoding (VLD) the variable length coded words from the bit stream to obtain locations and quantizer values of the non-zero DCT coefficients for each block. With the reconstruction of all non-zero DCT coefficients belonging to one block and subsequent inverse DCT, the quantized block pixel values are obtained. By processing the entire bit stream all picture blocks are decoded and reconstructed.

For coding P-pictures, the previously I- or P-picture picture *N-1* is stored in a picture store in both encoder and decoder. Motion compensation (MC) is performed on a Macroblock basis - only one motion vector is estimated between picture *N* and picture *N-1* for a particular Macroblock to be encoded. These motion vectors are coded and transmitted to the receiver. The motion compensated prediction error is calculated by subtracting each pel in a Macroblock with its motion shifted counterpart in the previous picture. A *8x8* DCT is then applied to each of the *8x8* blocks contained in the Macroblock followed by quantization of the DCT coefficients with subsequent run-length coding and entropy coding (VLC). A video buffer (VB) is needed to ensure that a constant target bit rate output is produced by the encoder. The quantization step-size can be adjusted for each Macroblock in a picture to achieve a given target bit rate and to avoid buffer overflow and underflow.

The decoder uses the reverse process to reproduce a Macroblock of picture *N* at the receiver. After decoding the variable length words (VLD) contained in the video decoder buffer (VB) the pixel values of the prediction error are reconstructed. The motion compensated pixels from the previous picture *N-1* contained in the picture store are added to the prediction error to recover the particular Macroblock of picture *N*.

An essential feature supported by the MPEG-1 coding algorithm is the possibility to update Macroblock information at the decoder only if needed - if the content of the Macroblock has changed in comparison to the content of the same Macroblock in the previous picture (Conditional Macroblock Replenishment). The key for efficient coding of video sequences at lower bit rates is the selection of appropriate prediction modes to achieve Conditional Replenishment. The MPEG standard distincts mainly between three different Macroblock coding types (MB types):

- *skipped MB* - prediction from previous picture with zero motion vector. No information about the Macroblock is coded nor transmitted to the receiver.

- *Inter MB* - motion compensated prediction from the previous picture is used. The MB type, the MB address and, if required, the motion vector, the DCT coefficients and quantization stepsize are transmitted.
- *Intra MB* - no prediction is used from the previous picture (Intra-picture prediction only). Only the MB type, the MB address and the DCT coefficients and quantization stepsize are transmitted to the receiver.

For accessing video from storage media the MPEG-1 video compression algorithm was designed to support important functionalities such as random access and fast forward (FF) and fast reverse (FR) playback functionalities. To incorporate the requirements for storage media and to further explore the significant advantages of motion compensation and motion interpolation, the concept of B-pictures (bi-directional predicted/bi-directional interpolated pictures) was introduced by MPEG-1. This concept is depicted in Fig. 8 for a group of consecutive pictures in a video sequence. Three types of pictures are considered: Intra-pictures (I-pictures) are coded without reference to other pictures contained in the video sequence. I-pictures allow access points for random access and FF/FR functionality in the bit stream but achieve only low compression. Inter-picture predicted pictures (P-pictures) are coded with reference to the nearest previously coded I-picture or P-picture, usually incorporating motion compensation to increase coding efficiency. Since P-pictures are usually used as reference for prediction for future or past pictures they provide no suitable access points for random access functionality or editability. Bi-directional predicted/interpolated pictures (B-pictures) require both past and future pictures as references. To achieve high compression, motion compensation can be employed based on the nearest past and future P-pictures or I-pictures. B-pictures themselves are never used as references.

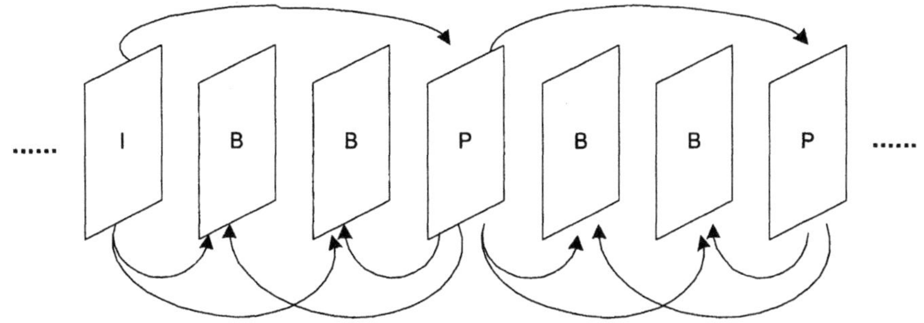

Figure 2.12 Sequence coded as I, P, and B pictures.

Fig. 2.12 shows I-pictures (I), P-pictures (P) and B-pictures (B) used in a MPEG-1 video sequence. B-pictures can be coded using motion compensated prediction based on the two nearest already coded pictures (either I-picture or P-picture). The arrangement of the picture coding types within the video sequence is flexible to suit the needs of diverse applications. The direction for prediction is indicated in the figure.

The encoder can configure the picture types in a video sequence with a high degree of flexibility to suit diverse applications requirements. As a general rule, a video sequence coded using I-pictures only (I I I I I I .....) allows the highest degree of random access, FF/FR and editability, but achieves only low compression. A sequence coded with a regular I-picture update and no B-pictures (i.e I P P P P P P I P P P P ...) achieves moderate compression and a certain degree of random access and FF/FR functionality. Incorporation of all three pictures types, as i.e. depicted in Fig. 2.12 (I B B P B B P B B I B B P ...), may achieve high compression and reasonable random access and FF/FR functionality but also increases the coding delay significantly. This delay may not be tolerable for two-way video communications, e.g. video-telephony or videoconferencing applications.

The standard video input format for MPEG-1 is non-interlaced. However, coding of interlaced colour television with both 525 and 625 lines at 29.97 and 25 pictures per second respectively is an important application for the MPEG-1 standard. A suggestion for coding ITU 601 digital color television signals has been made by MPEG-1 based on the conversion of the interlaced source to a progressive intermediate format. In essence, only one horizontally sub-sampled field of each interlaced video input picture is encoded, i.e. the sub-sampled top field. At the receiver the even field is predicted from the decoded and horizontally interpolated odd field for display. The necessary pre-processing steps required prior to encoding and the post-processing required after decoding are described in detail in the Informative Annex of the MPEG-1 specification [2-16].

MPEG-2 [2-17] MPEG-1 is an important and successful video coding standard with an increasing number of products becoming available on the market. The generic structure of the MPEG-1 supports a broad range of applications and applications specific parameters. However, there are needs for other standards to provide a video coding solution for applications not originally covered or envisaged by the MPEG-1 standard. Specifically, MPEG-2 was given the charter to provide video quality not lower than

NTSC/PAL and up to CCIR 601 quality. Emerging applications, such as digital cable TV distribution, networked database services via ATM, digital VTR applications and satellite and terrestrial digital broadcasting distribution, were seen to benefit from the increased quality expected to result from the new MPEG-2 standardization. MPEG-2 work was carried out in collaboration with the ITU-T SG 15 Experts Group for ATM Video Coding and in 1994 the MPEG-2 International Standard (which is identical to the ITU-T H.262 recommendation) was released. The specification of the standard is intended to be generic - hence the standard aims to facilitate the bit stream interchange among different applications, transmission and storage media.

Basically MPEG-2 can be seen as a superset of the MPEG-1 coding standard and was designed to be backward compatible to MPEG-1 - every MPEG-2 compatible decoder can decode a valid MPEG-1 bit stream. Many video coding algorithms were integrated into a single syntax to meet the diverse applications requirements. New coding features were added by MPEG-2 to achieve sufficient functionality and quality, thus prediction modes were developed to support efficient coding of interlaced video. In addition scalable video coding extensions were introduced to provide additional functionality, such as embedded coding of digital TV and HDTV, and graceful quality degradation in the presence of transmission errors.

However, implementation of the full syntax may not be practical for most applications. MPEG-2 has introduced the concept of "Profiles" and "Levels" to stipulate conformance between equipment not supporting the full implementation. Profiles and Levels provide means for defining subsets of the syntax and thus the decoder capabilities required to decode a particular bit stream.

As a general rule, each Profile defines a new set of algorithms added as a superset to the algorithms in the Profile below. A Level specifies the range of the parameters that are supported by the implementation (i.e. picture size, picture rate and bit rates). The MPEG-2 core algorithm at main profile (MP) features non-scalable coding of both progressive and interlaced video sources. It is expected that most MPEG-2 implementations will at least conform to the MP at main level (ML), also represented as MP@ML, which supports non-scalable coding of digital video with approximately digital TV parameters - a maximum sample density of 720 samples per line and 576 lines per picture, a maximum picture rate of 30 pictures per second and a maximum bit rate of 15 Mbit/s.

The MPEG-2 algorithm defined in the MP is a straightforward extension of the MPEG-1 coding scheme to accommodate coding of interlaced video, while retaining the full range of functionality provided by MPEG-1. Identical to the MPEG-1 standard, the MPEG-2 coding algorithm is based on the general Hybrid DCT/DPCM coding scheme as outlined in Fig. 2.5, incorporating a Macroblock structure, motion compensation and coding modes for conditional replenishment of Macroblocks. The concept of I-pictures, P-pictures and B-pictures as introduced in Fig. 2.12 is fully retained in MPEG-2 to achieve efficient motion prediction and to assist random access functionality. Notice that the algorithm defined with the MPEG-2 SIMPLE Profile is basically identical with the one in the MP, except that no B-picture prediction modes are allowed at the encoder. Thus the additional implementation complexity and the additional picture stores necessary for the decoding of B-pictures are not required for MPEG-2 decoders only conforming to the Simple Profile.

Field and Frame Pictures: MPEG-2 has introduced the concept of *frame pictures* and *field pictures* along with particular *frame prediction* and *field prediction* modes to accommodate coding of progressive and interlaced video. For interlaced sequences it is assumed that the coder input consists of a series of odd (top) and even (bottom) fields that are separated in time by a field period. Two fields of a Frame may be coded separately. In this case each field is separated into adjacent non-overlapping Macroblocks and the DCT is applied on a field basis. Alternatively two fields may be coded together as a frame (frame pictures) similar to conventional coding of progressive video sequences. Here, consecutive lines of top and bottom fields are simply merged to form a frame. Notice, that both frame pictures and field pictures can be used in a single video sequence.

The concept of field-picture prediction can be explained briefly as follows. The top fields and the bottom fields are coded separately. However, each bottom field is coded using motion compensated Inter-field prediction based on the previously coded top field. The top fields are coded using motion compensated Inter-field prediction based on either the previously coded top field or based on the previously coded bottom field. This concept can be extended to incorporate B-pictures.

Field and Frame Prediction: New motion compensated field prediction modes were introduced by MPEG-2 to efficiently encode field pictures and frame pictures. In field prediction, predictions are made independently for each field by using data from one or more previously decoded field, i.e. for a top field a prediction may be obtained from either a previously decoded top field (using motion compensated prediction) or from the previously decoded bottom field belonging to the same frame. Generally the Inter-field prediction from the decoded field in the same frame is preferred if no motion occurs between fields. An indication which reference field is used for prediction is transmitted with the bit stream. Within a field picture all predictions are field predictions.

Frame prediction forms a prediction for a frame picture based on one or more previously decoded frames. In a frame picture either field or frame predictions may be used and the particular prediction mode preferred can be selected on a Macroblock-by-Macroblock basis. It must be understood, however, that the fields and frames from which predictions are made may have themselves been decoded as either field or frame pictures.

MPEG-2 also has introduced new motion compensation modes to efficiently explore temporal redundancies between fields, namely the "Dual Prime" prediction and the motion compensation based on 16x8 blocks.

Chrominance Formats: MPEG-2 has specified additional Y:Cb:Cr luminance and chrominance sub-sampling ratio formats to assist applications with highest video quality requirements. Next to the 4:2:0 format already supported by MPEG-1 the specification of MPEG-2 is extended to 4:2:2 formats as the 422 Profile that is suitable for studio video coding applications.

MPEG-4 [2-18] Compared to MPEG-1 and MPEG-2, the MPEG-4 standard brings a new paradigm as it treats a scene to be coded as consisting of individual objects; thus each object in the scene can be coded individually and the decoded objects can be composed in a scene. MPEG-4 is optimized [2-19,2-20] for bit-rate range of 10 kbit/s to 3 Mbit/s. The work done by ITU-T for H.263 version 2 [2-23] is of relevance for MPEG-4 since H.263 version 2 is an extension of H.263 [2-24], and since H.263 was also one of the starting basis for MPEG-4. However, MPEG-4 is a more complete standard [2-25] due to its ability to address a very wide range and types of applications, extensive systems support, and tools for coding and integration of natural and synthetic objects.

An input video sequence consists of a related snapshots or pictures, separated in time. Each picture consists of temporal instances of objects that undergo a variety of changes such as translations, rotations, scaling, brightness and color variations etc. Moreover, new objects enter a scene and/or existing objects depart, resulting in appearance of certain objects only in certain pictures. Sometimes, scene change occurs, and thus the entire scene may either get reorganized or replaced by a new scene. Many of MPEG-4 functionalities require access not only to entire sequence of pictures, but to an entire object, and further, not only to individual pictures, but also to temporal instances of these objects within a picture. A temporal instance of a video object can be thought of as a snapshot of an arbitrary shaped object that occurs within a picture, such that like a picture, it is intended to be an access unit, and, unlike a picture, it is expected to have a semantic meaning.

The concept of Video Objects and their temporal instances, Video Object Planes (VOPs) is central to MPEG-4 video. A VOP can be fully described by texture variations (a set of luminance and chrominance values) and (explicit or implicit) shape representation. In natural scenes, VOPs are obtained by semi-automatic or automatic segmentation, and the resulting shape information can be represented as a *binary shape* mask. On the other hand, for hybrid (of natural and synthetic) scenes generated by blue screen composition, shape information is represented by an 8-bit component, referred to as *gray scale shape*. Video Objects (VOs) can also be subdivided into multiple representations or Video Object Layers (VOLs), allowing scalable representations of the video object. If the entire scene is considered as one object and all VOPs are rectangular and of the same size as each picture then a VOP is identical to a picture. Additionally, an optional Group of Video Object Planes (GOV) can be added to the video coding structure to assist in random access operations.

Fig. 2.13 shows the decomposition of a picture into a number of separate VOPs. The scene consists of two objects (head of a lion, and a logo) and the background. The objects are segmented by semi-automatic or automatic means and are referred to as VOP1 and VOP2, while the background (the gray area) without the two objects is referred to as VOP0. Each picture in the sequence is segmented into VOPs in this manner. Thus, a segmented sequence contains a temporal set of VOP0's, a temporal set of VOP1's and a temporal set of VOP2's.

Each of the VOs are coded separately and multiplexed to form a bitstream that users can access and manipulate (cut, paste,..). The encoder sends together with video objects, information about scene composition to indicate where and when VOPs of a video object are to be displayed. This information is however optional and may be ignored at the decoder which may use user specified information about composition.

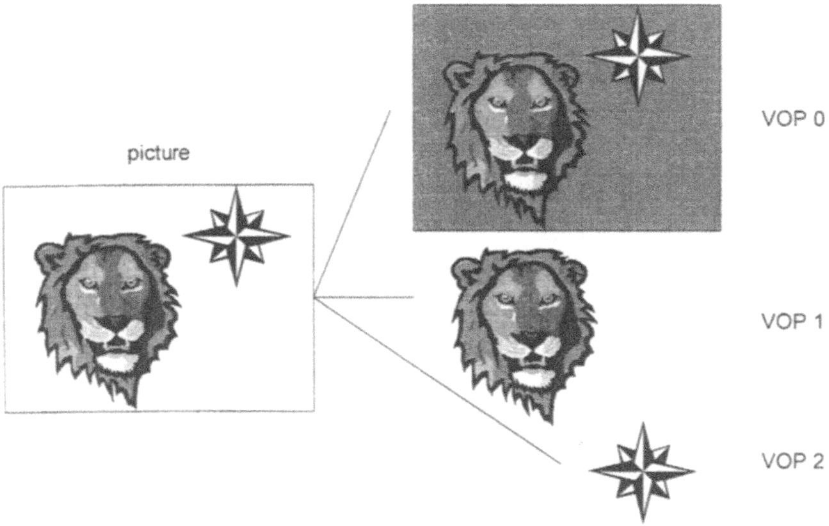

Figure 2.13 Semantic segmentation of a picture in to VOPs

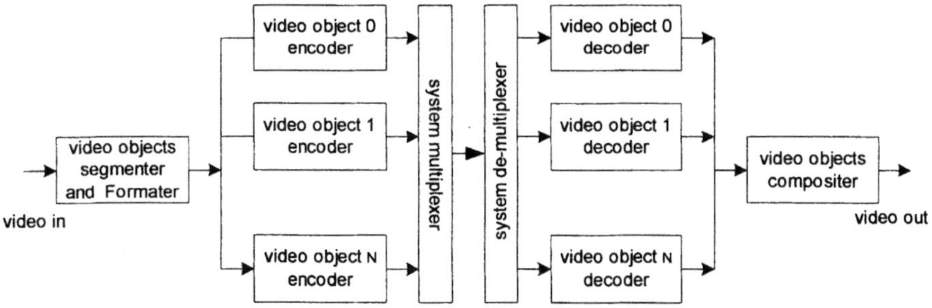

Figure 2.14 A block diagram of Video Object based codec of MPEG-4 video

In Fig. 2.14, a high level logical structure of a video object based coder is shown. Its main components are Video Objects Segmenter/Formatter, Video Object Encoder, Systems Multiplexer Systems Demultiplexer, Video Object Decoder and Video Object Compositor. Video Object Segmenter segments the input scene into video objects for encoding by Video Object Encoder. The coded data of various video objects is multiplexed for storage or transmission, following which it is demultiplexed and decoded by video object decoders and offered to compositer, which renders the decoded scene.

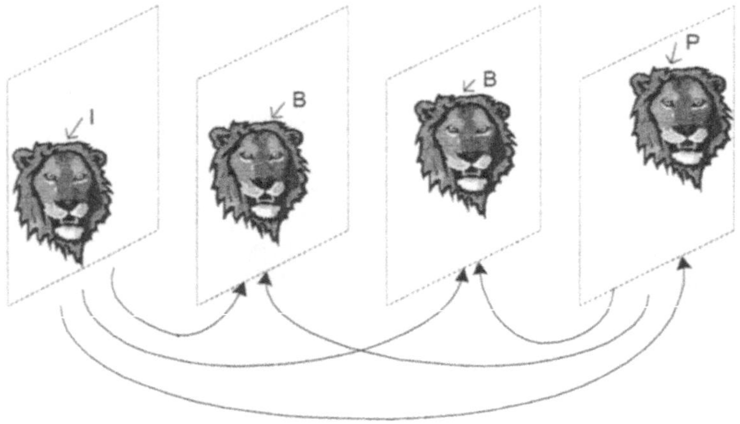

Figure 2.15 An example prediction structure when using I, P and B-VOPs

To consider how coding takes place in a video object encoder, consider a sequence of VOPs. MPEG-4 video extends the concept of intra (I-) pictures, predictive (P-) and bidirectionally predictive (B-) pictures of MPEG-1/2 video to VOPs, thus I-VOP, P-VOP and B-VOP result. Fig. 2.15 shows a coding structure which uses two consecutive B-VOPs between a pair of reference VOPs (I- or P-VOPs).

The basic MPEG-4 coding employs motion compensation and (8x8) DCT based coding and shape coding. Each VOP is comprised of macroblocks that can be coded as intra- or as inter- macroblocks.The definition of a macroblock is exactly the same as in MPEG-1 and MPEG-2. In I-VOPs, only intra-macroblocks exist. In P-VOPs, intra as well as unidirectionally predicted macroblocks can occur where as in B-VOPs, both uni- or bidirectionally predicted- macroblocks can occur. The gray level shape (alpha) is coded as Y component of the video while binary shape (alpha) is coded by using an integer arithmetic-coding algorithm [2-19][2-20].

MPEG-4 has made several improvements in coding of intra macroblocks (INTRA) as compared to H.263, MPEG-1/2. In particular it supports the following:

- DPCM prediction of the DC coefficient [2-25],
- DPCM prediction of a subset of AC coefficients [2-25],
- Specialized coefficient scanning based on the coefficient prediction,
- Huffman table selection,
- Non-Linear inverse DC Quantization.

As in the previous MPEG standards, inter-macroblocks in P- and B- VOPs are coded using a motion compensated block matching technique to determine the prediction error. However, because a VOP is arbitrarily shaped, and the size can change from one instance to the next, special padding techniques are defined to maintain the integrity of the motion compensation. For this process, the minimum bounding rectangle of each VOP is referenced to an absolute frame coordinate system. All displacements are with respect to the absolute coordinate system so that no VOP alignment is necessary. Enhanced motion compensation options are developed in MPEG-4:

- Direct-mode bi-directional prediction [2-26][2-27][2-28],
- Quarter-pixel motion compensation,
- Global motion compensation techniques,

Neither the H.263 nor the MPEG-1 standard allows a separate variable length Huffman code (VLC) table for coding DCT coefficients of intra blocks. This forces the use of the inter block DCT VLC table which is inefficient for intra blocks. The MPEG-2 standard does allow a separate VLC table for intra blocks but it is optimized for much higher bit-rates. MPEG-4 provides an additional table optimized for coding of AC coefficients of intra blocks [2-19]. The MPEG-4 table is 3 dimensional; that is it maps the zero run length, the coefficient level value, and the last coefficient indication into the variable length code.

Rate Control: An important feature supported by the MEPG video encoding algorithms is the possibility to tailor the bitrate (and thus the quality of the reconstructed video) to specific applications requirements by adjusting the quantizer step-size of the quantization block in Fig. 2.16 for quantizing the DCT-coefficients. Coarse quantization of the DCT-coefficients enables the storage or transmission of video with high compression ratios, but, depending on the level of quantization, may result in significant coding artifacts. The MPEG video standards allow the encoder to select different quantizer values for each coded Macroblock - this enables a high degree of flexibility to allocate bits in pictures where needed to improve picture quality. Furthermore it allows the generation of both constant and variable bit-rates for storage or real-time transmission of the compressed video.

Compressed video information is inherently variable in nature. This is caused by the, in general, variable content of successive video pictures. To store or transmit video at constant bit rate it is therefore necessary to buffer the bitstream generated in the encoder in a video buffer (VB) as depicted in Fig. 2.16. The input into the encoder VB is variable over time and the output is a constant bitstream. At the decoder the VB input bitstream is constant and the output used for decoding is variable. MPEG encoders and decoders implement buffers of the same size to avoid reconstruction errors.

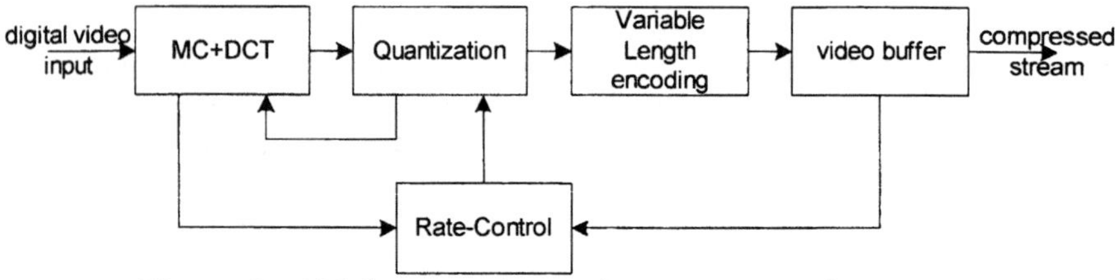

Figure 2.16 Video encoder with a rate-control engine

A rate control algorithm at the encoder adjusts the quantizer step-size depending on the video content and activity to ensure that the video buffers will never overflow - while at the same time targeting to keep the buffers as full as possible to maximize picture quality. In theory overflow of buffers can always be avoided by using a large enough video buffer. However, besides the possibly undesirable costs for the implementation of large buffers, there may be additional disadvantages for applications requiring low-delay between encoder and decoder, such as for the real-time transmission of conversational video. If the encoder bitstream is smoothed using a video buffer to generate a constant bit rate output, a delay is introduced between the encoding process and the time the video can be reconstructed at the decoder. Usually the larger the buffer the larger the delay introduced.

MPEG has defined a minimum video buffer size that needs to be supported by all decoder implementations. This value also determines the maximum value of the VB size that an encoder needs to use for generating a bitstream. However, to reduce delay or encoder complexity, it is possible to choose a virtual buffer size value at the encoder smaller than the minimum VB size which needs to be supported by the decoder. This virtual buffer size value is transmitted to the decoder before sending the video bitstream. The detailed discussion on video buffer is given in Chapters 3 and 6.

The rate control algorithm used to compress video is not part of the MPEG standards and it is thus left to the implementers to develop efficient strategies. It is worth emphasizing that the efficiency of the rate control algorithms selected by manufacturers to compress video at a given bit rate heavily impacts on the visible quality of the video reconstructed at the decoder.

Currently, there are a number of standard-based video compression technologies that are applied in various digital video services. For example, the standards discussed in this section are MPEG-1, MPEG-2, MPEG-4, Motion JPEG, H.261 and H.263. Digital compression can take these many forms and be suited to a multitude of applications. Each compression scheme has its strengths and weaknesses because the codecs you choose will determine how good the images will look and how smoothly the images will flow.

As one looks towards the future, it seems clear that more advanced video compression standards (e.g. MPEG-4 part-10, also called H.26L) are destined to replace existing standards in many applications (e.g. video streaming) that require a lower bit rate. Also, the need for higher compression efficiency in many commercial systems, such as video on demand and satellite broadcasting digital video, seems certain to spur a continuing interest in the design of extremely powerful compression algorithms. Finally, the technical challenges inherent in designing new compression systems will continue to lead to further advances in digital video communications.

Bibliography

[2-1] N. S. Jayant and P. Noll, Digital Coding of Waveform, Englewood, Cliffs, NJ: Prentice-Hall, 1984.

[2-2] N.Ahmed, T.Natrajan and K.R.Rao, "Discrete Cosine Transform", IEEE Trans. on Computers, Vol. C-23, No.1, pp. 90-93, December 1984.

[2-3] A. K. Jain, Fundamentals of Digital Image Processing, Englewood Cliffs, NJ: Prentice-Hall, 1989.

[2-4] W. B. Pennebaker and J. L. Mitchell, JPEG Still Image Data Compression Standard, New York: Van Nostrand Reinhold, 1993.

[2-5] J. W. Woods (ed.), Subband Image Coding, Boston: Kluwer Academic Publishers, 1991.

[2-6] K. Jack, Video Demystified, 3rd ed., San Diego: HighText Interactive, 2000.

[2-7] T. M. Cover and J. A. Thomas, Elements of Information Theory, New York: Wiley, 1991.

[2-8] B. G. Haskell, A. Puri, and A. N. Netravali, Digital Video: An Introduction to MPEG-2, New York: Chapman & Hall, 1997.

[2-9] A. Gersho and R. M. Gray, Vector Quantization and Signal Compression, Boston: Kluwer Academic Publishers, 1992.

[2-10] Xuemin Chen, article "Data compression for networking", Wiley Encyclopedia of Electrical and Electronics Engineering, Vol.4, pp.675-686, 1999.

[2-11] H. S. Malvar, Signal Processing with Lapped Transforms. Boston: Artech House, 1992, Chapter 2.

[2-12] K. R. Rao and P. Yip. Discrete Cosine Transform: *Algorithms, Advantages, Application*, Boston: Academic Press, 1990, Chapter 4.

[2-13] W. Cham, "Development of integer cosine transforms by the principle of dyadic symmetry," *IEE Proc.*, Part 1, vol. 136, pp. 276–282, Aug. 1989.

[2-14] R.Schäfer and T.Sikora, "Digital Video Coding Standards and Their Role in Video Communications", Proceedings of the IEEE Vol. 83, pp. 907-923, 1995.

[2-15] T.Sikora, "The MPEG-1 and MPEG-2 Digital Video Coding Standards", IEEE Signal Processing Magazine, .

[2-16] ISO/IEC 11172-2, "Information Technology - Coding of Moving Pictures and Associated Audio for Digital Storage Media at up to about 1,5 Mbit/s - Video", Geneva, 1993

[2-17] ISO/IEC 13818, "Information Technology - Generic Coding of Moving Pictures and Associated Audio, Recommendation H.262", International Standard, Paris, 25 March 1995.

[2-18] ISO/IEC 14496-2, Information Technology – Generic coding of audio-visual objects – Part 2: Visual, Atlantic City, Nov. 1998

[2-19] Atul Puri and T. H. Chen, Multimedia Standards and Systems, Chapman & Hall, New York, 1999.

[2-20] Krit Panusopone, Xuemin. Chen, B. Eifrig and Ajay. Luthra, "Coding tools in MPEG-4 for interlaced video," IEEE Transactions on circuits and systems for video technology, vol. No. , Apr. 2000.

[2-21] T.Sikora, "The MPEG-4 Video Standard Verification Model," IEEE Transactions on circuits and systems for video technology, Vol.7, No.1, Feb.1997.

[2-22] R. Talluri, "Error Resilient Video Coding in ISO MPEG-4 Standard," IEEE Communications Magazine, June 1998

[2-23] ITU-T Experts Group on Very Low Bitrate Visual Telephony, "ITU-T Recommendation H.263 Version 2: Video Coding for Low Bitrate Communication," Jan. 1998.

[2-24] ITU-T Experts Group on Very Low Bitrate Visual Telephony, "ITU-T Recommendation H.263: Video Coding for Low Bitrate Communication," Dec. 1995.

[2-25] Robert O. Eifrig, Xuemin Chen, and Ajay Luthra, "Intra-macroblock DC and AC coefficient prediction for interlaced digital video", US Patent Number 5974184, Assignee: General Instrument Corporation, Oct. 26, 1999.
[2-26] Robert O. Eifrig, Xuemin Chen, and Ajay Luthra, "Prediction and coding of bi-directionally predicted video object planes for interlaced digital video", US Patent Number 5991447, Assignee: General Instrument Corporation, Nov. 23, 1999.
[2-27] Robert O. Eifrig, Xuemin Chen, and Ajay Luthra, "Motion estimation and compensation of video object planes for interlaced digital video", US Patent Number 6005980, Assignee: General Instrument Corporation, Dec. 21, 1999.
[2-28] Robert O. Eifrig, Xuemin Chen, and Ajay Luthra, "Motion estimation and compensation of video object planes for interlaced digital video", US Patent Number 6026195, Assignee: General Instrument Corporation, Feb. 15, 2000.
[2-29] Xuemin Chen, Robert O. Eifrig, Ajay Luthra, and Krit Panusopone, "Coding of an arbitrarily shaped interlaced video in MPEG-4", IEEE International Conference on Acoustics, Speech, and Signal Processing, Vol. 6, pp.3121-3124, 1999.
[2-30] Krit Panusopone and Xuemin Chen, "A fast motion estimation method for MPEG-4 arbitrarily shaped objects", IEEE International Conference on Image Processing, Vol. 3, pp.624-627, 2000.

3 Buffer Constraints on Compressed Digital Video

3.1 Video Compression Buffers

In this chapter, constraints on video compression/decompression buffers and the bit rate of a compressed video bit stream are discussed. These constraints are imposed by the transmission channels. First, concepts of compressed video buffers are introduced. Then, conditions that prevent the video encoder and decoder buffer overflow or underflow are derived for the channel that can transmit a variable bit rate video. Next, strategies for buffer management are developed from these derived conditions. Examples are given to illustrate how these buffer management ideas can be applied in a compression system that controls both the encoded and transmitted bit rates. Buffer verification problem for channels with rate constraints, e.g. constant-rate and leaky-bucket channels, is also discussed.

As discussed in Chapter 1, uncompressed video is constant rate by nature and is transmitted over constant-rate channels, e.g. analog TV signal over terrestrial and cable broadcasting networks. For transmission of compressed digital video, since most video compression algorithms use variable length codes, e.g. Huffman codes, a buffer at the encoder is necessary to translate the variable rate output from the video compression engine into the constant-rate channel. A similar buffer is also necessary at the decoder to convert the constant channel bit-rate stream into a variable bit-rate streams for decompression.

In general case, compressed video can also be transmitted over variable-rate channels, e.g. statistically multiplexed (often called StatMux) transport channels and broadband IP networks. These networks are able to support variable bit rates by partitioning video data into a sequence of packets and inputting them to the network asynchronously. In another words, these networks may allow video to be transmitted on a channel with variable rate.

Recently, some broadband networks, such as StatMux DTV channel, Asynchronous Transfer Mode (ATM) network and high-speed Ethernet, are deployed for transmitting video because they can accommodate the bit rate necessary for high-quality video, and also because the quality of the video can benefit from the variable bit rate that these networks can provide. As a result, video compression algorithms can have less-constrained bit rates to achieve constant quality. The algorithm designed for a variable-rate channel is usually more efficient than the algorithm designed for a constant-rate channel [3-1].

However, if the bit rate of coded streams is allowed to vary arbitrarily, the network will be unable to provide guaranteed delivery of all packets in real time. There are two solutions to overcome this problem [3-2]. The first solution is to have the user assign a priority (e.g. high or low) to each packet transmitted to the network. The high-priority packets are almost guaranteed by the network for delivering while the low-priority packets can be dropped by the network. The second solution, which is additional to the first one, is to assume that a contract exists between the network and the user. The network ensures that the cell-loss rate (CLR) for high-priority packets will not exceed a certain value. A policing function monitors the user output and either drops packets in excess of the contract or marks these excess packets as low priority, possibly to be dropped later in the network.

The advantage of priority labeling for both video and network have been well established [3-3] - [3-7]. In addition, the effect of a policing function on the network behavior has also been studied [3-7] - [3-11]. The existence of a policing function has a significant effect on video transmission because certain information are essential to the decoder, e.g. timing information, various start codes, etc. If this information is not received, the video decoder will be unable to decode any pictures. Therefore, it is very important to the video user that all high-priority packets are received. This implies that the network should never drop high-priority packets or, equivalently, that the network should never change the marking of a high-priority packet to low priority. Therefore, it is important that the video algorithm can control its output bit rate to ensure that the network-imposed policing function does not detect any excess high-priority packets.

3.2 Buffer Constraints for Variable-Rate Channels

It is shown in this section that for a constant-rate channel, it is possible to prevent the decoder buffer from over-flowing or under-flowing simply by ensuring that the encoder buffer never underflows or overflows.

For a variable-rate channel, additional constraints must be imposed on the encoding rate, the channel rate, or both. This section also examines the constraints imposed on the encoded video bit-rate as a result of encoder and decoder buffering.

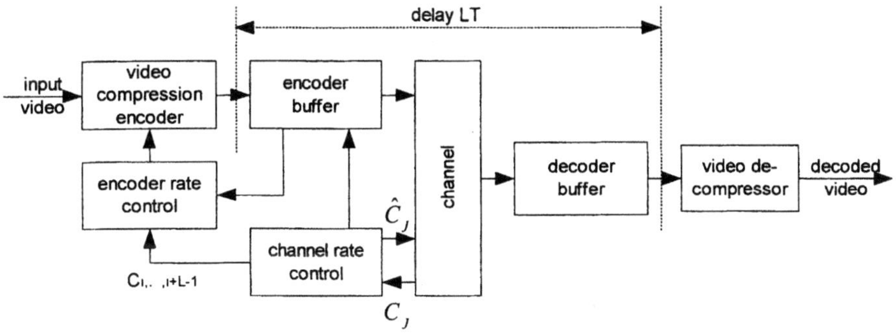

Figure 3.1 A simplified model of the encoder-decoder pair with buffers.

Figure 3.1 shows a model of video compression and decompression engines with the corresponding rate-control devices and buffers.

Intuitively, if either encoder or decoder buffer overflows, information is lost. Encoder buffer underflow is only a problem if the channel has constant bit rate and cannot be turned off. In this case, some non-video data must be transmitted. Since encoder buffer underflow can always be avoided by sending stuffing bits, it is not considered as a problem.

However, the concept of decoder buffer underflow is less intuitive since the real-time decoder is generally capable of removing bits from its buffer faster that bits arrive. The decoder buffer is said to underflow when the decoder must display a new picture, but no new picture has finished decoding.

Therefore, this is the case that the following situations are happened simultaneously:

- The decoder video buffer is empty.
- The picture display buffer is not ready (full).
- It is time to display a newly decoded picture.

For a constant bit-rate channel, it is possible to determine upper bounds on encoder and decoder buffer sizes such that if the encoder's output rate is controlled to ensure no encoder buffer overflow or underflow, then the decoder buffer will also never underflow or overflow. However, as one will see, the situation becomes more complex when the channel may transmit a variable bit-rate stream, for example, when transmitting video across packet switched networks. These upper bounds on buffer sizes are pre-determined both in terms of a constraint on the encoder rate and a constraint on the channel rate. The channel rate may be variable but is not otherwise constrained. Many researches on this topic have been reported [3-2]-[3-5]. To understand the constraints on buffer sizes, one first needs to analyze the buffer dynamics.

3.2.1 Buffer Dynamics

The encoder and decoder buffer dynamics can be characterized by the following defined parameters. Define $E_O(t)$ to be the number of units (e.g. bits, bytes or packets) output by the encoder at time *t*. The channel bit rate $R_C(t)$ is variable. $B_E(t)$ and $B_D(t)$ are the fullness of the encoder and decoder buffers at time *t*, respectively. Each buffer has a maximum size, $B_E^{\max}$ and $B_D^{\max}$, that cannot be exceeded. Given $B_E^{\max}$, the encoder is designed to ensure its buffer never overflows, i.e.,

$$0 \le B_E(t) \le B_E^{\max}, \qquad \text{for all } t. \tag{3.1}$$

Define $S(iT)$ (i =1,2,......) to be the number of units in the interval $[(i-1)T, iT]$, where is the picture duration of the original uncompressed video, e.g. $T = 1/29.97$ second for digitized NTSC video. Therefore,

$$S(iT) = \int_{(i-1)T}^{iT} E_O(t)dt. \tag{3.2}$$

Similarly, let $C(iT)$ be the number of bits that are transmitted during the i-th picture period:

$$C(iT) = \int_{(i-1)T}^{iT} R_C(t)dt\,. \tag{3.3}$$

The encoder buffer receives bit at rate $E_O(t)$ and outputs bits at rate $R_C(t)$. Therefore, assuming empty buffers prior to startup at time $t = 0$

$$B_E(t) = \int_0^t \left[E_O(u) - R_C(u)\right] du\,, \tag{3.4}$$

and the encoder buffer fullness after encoding picture i is

$$B_E^i(T) = B_E(iT) = \int_0^{iT} \left[E_O(u) - R_C(u)\right] du\,. \tag{3.5}$$

This can also be written as

$$B_E^i(T) = \sum_{j=1}^{i} S(jT) - \sum_{j-1}^{i} C(jT), \tag{3.6}$$

or recursively as

$$B_E^i(T) = B_E^{i-1}(T) + S(iT) - C(iT)\,. \tag{3.7}$$

After the decoder begins to receive the compressed stream, it waits $L \cdot T$ before starting to decode. For clarity, let us assume that L is an integer, although this is not necessary. At the decoder, define a decoding time index τ, which is zero when decoding starts.

$$t = \tau + L \cdot T + \Delta(\tau), \tag{3.8}$$

where $\Delta(\tau)$ denotes the channel delay at τ.

Next, conditions on the buffers and the channel are examined to ensure the decoder buffer never overflows or underflows, i.e.,

$$0 \le B_D(\tau) \le B_D^{\max}, \qquad \text{for all } \tau\,. \tag{3.9}$$

The initial decoder-buffer fullness $B_D(0)$ can be calculated by the encoder if L is predetermined or sent explicitly as a decoder parameter. It is given by

$$B_D^0(T) = \sum_{j=1}^{L} C(jT). \tag{3.10}$$

The decoder buffer fullness at time $\tau = iT$ is then given by

$$B_D^i(T) = B_D^{i-1}(T) + C((L+i) \cdot T) - S(i \cdot T), \tag{3.11}$$

$$B_D^i(T) = B_D^0(T) + \sum_{j=1}^{i} C((L+j) \cdot T) - \sum_{j=1}^{i} S(j \cdot T)\,. \tag{3.12}$$

For $(i-1)T < \tau < iT$, the decoder buffer fullness varies, depending on the channel rate $R_C(t)$ and the rate at which the decoder extracts data from its buffer. In this time interval, the decoder buffer fullness could increase to the higher levels of $B_D^{i-1}(T)+C((i-1)\cdot T)$ or $B_D^i(T)+C(i\cdot T)$, or decrease to the lower level of $B_D^{i-1}(T)-C(i\cdot T)$ or $B_D^i(T)-C((i+1)\cdot T)$.

There are two useful expressions for $B_D^i = B_D^i(T)$ when the channel has variable rate, each derived using Eqs. (3.6), (3.10) and (3.12).

$$\begin{aligned} B_D^i &= \sum_{j=1}^{L} C(jT) + \sum_{j=L+1}^{L+i} C(jT) - \sum_{j=1}^{i} S(jT) \\ &= \sum_{j=i+1}^{i+L} C(jT) - \left(\sum_{j=1}^{i} S(jT) - \sum_{j=1}^{i} C(jT) \right) \\ &= \sum_{j=i+1}^{i+L} C(jT) - B_E^i(T), \end{aligned} \tag{3.13}$$

or

$$\begin{aligned} B_D^i &= \sum_{j=i+1}^{i+L} S(jT) - \left(\sum_{j=1}^{i+L} S(jT) - \sum_{j=1}^{i+L} C(jT) \right) \\ &= \sum_{j=i+1}^{i+L} S(jT) - B_E^{i+L}(T). \end{aligned} \tag{3.14}$$

It is shown in Eq. (3.13) that $B_D^i(T)$ is a function of the cumulative channel rates over the last L pictures and the encoder buffer fullness L pictures ago, when picture i was encoded. In Eq. (3.14), $B_D^i(T)$ can also be expressed as a function of the cumulative encoder rates over the last L pictures and the encoder buffer fullness now, or when picture $i+L$ is encoded. Eq. (3.14) is an expression that the encoder can compute directly from its observations.

3.2.2 Buffer Constraints

Next, conditions necessary to prevent encoder and decoder buffer underflow and overflow are derived for a variable-rate channel. Eqs. (3.1) and (3.7) yield the following conditions for preventing encoder buffer overflow and underflow:

$$0 \le B_E^i(T) = B_E^{i-1}(T) + S(iT) - C(iT) \le B_E^{\max}, \tag{3.15}$$

$$C(iT) - B_E^{i-1}(T) \le S(iT) \le C(iT) + B_E^{\max} - B_E^{i-1}(T), \tag{3.16}$$

which is a constraint on the number of bits per coded picture for a given channel rate. For example, when the channel has a constant rate, the encoder prevents its buffer from over-flowing or under-flowing by varying the quality of coding [3-13]. If the encoder is informed that its buffer is too full, it reduces the bit rate being input to the buffer by reducing the quality of coding, e.g. using a coarser quantizer on the DCT coefficients. Conversely, if encoder buffer underflow threatens, the encoder can generate more input data, either by increasing the quality of coding or by outputting stuffing data that are consistent with the specified coding syntax.

Alternatively, to achieve constant picture quality, one can instead let the number of bits per picture $S(iT)$ be unconstrained, and force the accumulated channel rate $C(iT)$ to accommodate. Rewriting Eq. (3.15), one has

$$-B_E^{\max} \le C(iT) - S(iT) - B_E^{i-1}(T) \le 0, \tag{3.17}$$

i.e.

$$S(iT) + B_E^{i-1}(T) - B_E^{\max} \le C(iT) \le S(iT) + B_E^{i-1}(T). \tag{3.18}$$

The left inequality provides the encoder underflow condition while the right inequality shows the encoder overflow condition.

Therefore, encoder buffer overflow and underflow can he prevented by constraining either the encoded bit rate per picture period given by Eq. (3.16) or the transmitted bit rate per picture period given by Eq. (3.18).

To prevent decoder buffer overflow and underflow, one can combine Eqs. (3.9) and (3.11) to obtain

$$0 \le C((i+L)\cdot T) + B_D^{i-1}(T) - S(i\cdot T) \le B_D^{\max}, \tag{3.19}$$

$$C((i+L)\cdot T) + B_D^{i-1}(T) - B_D^{\max} \le S(i\cdot T) \le C((i+L)\cdot T) + B_D^{i-1}(T), \tag{3.20}$$

which is a constraint on the encoder bit rate for a given channel rate.

Alternatively, one can again allow the number of bits per picture to be unconstrained and examine the constraint on the channel rate $C(iT)$.

$$S(iT) - B_D^{i-1}(T) \le C((i+L)T) \le S(iT) + (B_D^{\max} - B_D^{i-1}(T)), \tag{3.21}$$

or, for $i > L$,

$$S((i-L)T) - B_D^{i-L-1}(T) \le C(iT) \le S((i-L)T) + \left(B_D^{\max} - B_D^{i-L-1}(T)\right) \tag{3.22}$$

decoder underflow condition — decoder overflow condition

This provides a restriction on the accumulated channel rate $C(iT)$ that depends on the encoder activity L pictures ago.

Even if the channel rate is completely controllable, a restriction still exists on $S(iT)$, the number of bits used to encode picture i. This constraint is necessary to prevent simultaneous overflow of both buffers. Note that simultaneous underflow of both buffers is not a problem. The upper bound of Eq.(3.18) is always greater than the lower bound of Eq. (3.21).

It can be seen either by combining the lower bound of Eq.(3.18) with the upper bound of Eq.(3.21),

$$S(iT)-\left(B_E^{\max}-B_E^{i-1}(T)\right)\leq C(iT)\leq S((i-L)T)+\left(B_D^{\max}-B_D^{i-L-1}(T)\right), \quad (3.23)$$

$$S(iT)\leq S((i-L)T)+\left(B_E^{\max}-B_E^{i-1}(T)\right)+\left(B_D^{\max}-B_D^{i-L-1}(T)\right) \quad (3.24)$$

or by noting that because the delay is LT, the system must store L pictures worth of data,

$$0\leq \sum_{j=i-L+1}^{i} S(iT)\leq B_E^{\max}+B_D^{\max}. \quad (3.25)$$

These bounds arise because of the finite memory of the video system. The system can store no more than $B_E^{\max}+B_D^{\max}$ bits at any given time, but it must always store L pictures of data. Therefore, these L pictures cannot be coded with too many bits. In the case of equality for either Eqs.(3.24) or (3.25), both buffers are completely full at the end of picture .

In the above discussion, the channel is assumed to have a constant delay. However, for many applications, e.g. video transmission over a packet switch network, the channel is expected to have a variable delay. To accommodate such variable delay, the largest possible channel delay should be used in Eq. (3.8). In addition, the decoder buffer should be large enough to contain the additional bits that may arrive with shorter delay. Thus if the minimum channel delay is $\Delta_{\min}$ and the maximum channel delay is $\Delta_{\max}$, Eq. (3.8) becomes

$$t=\tau+L\cdot T+\Delta_{\max} \quad (3.26)$$

and the decoder buffer constraint of Eq. (3.19) becomes

$$0\leq B_D^{i-1}(T)+C((i+L)T)-S(iT)\leq B_D^{\max}-\max_{i}\int_{iT}^{iT+\Delta_{\max}-\Delta_{\min}} R_C(u)\,du. \quad (3.27)$$

3.3 Buffer Verification for Channels with Rate Constraints

3.3.1 Constant-Rate Channel

If the channel has a constant bit rate, then the buffer verification problem can be simplified. In particular, it is possible to ensure that the decoder buffer never overflows or underflows, provided that the encoder buffer never overflows or underflows. For the constant-rate channel let $C(iT) = R_C \cdot T$ be the number of bits transmitted during one uncompressed picture period of duration T. The initial fullness of the decoder buffer when decoding starts is

$$B_D(0) = B_D^0(T) = L \cdot T \cdot R_C . \tag{3.28}$$

Eq.(3.12) can be simplified as

$$B_D^J(T) = B_D^0(T) - B_E^J(T), \tag{3.29}$$

for the channel that has a constant rate. Note that this equation is not true for a variable-rate channel since, in that case,

$$\begin{aligned} B_D^J(T) &= B_D^0(T) + \sum_{i=1}^{J} C((L+i) \cdot T) - \sum_{i=1}^{J} S(i \cdot T) \\ &\neq B_D^0(T) + \sum_{i=1}^{J} C(i \cdot T) - \sum_{i=1}^{J} S(i \cdot T) \\ &= B_D^0(T) - B_E^J(T). \end{aligned} \tag{3.30}$$

Because $B_E^J(T)$ is always non-negative, the decoder buffer is never as full at the end of a picture as it was before decoding started. Therefore, to prevent decoder buffer overflow, using Eq.(3.29), the decoder buffer size can be chosen solely to ensure that it can handle the initial buffer fullness, plus the number of bits for one picture. In most cases, the decoder is much faster than the channel rate. Thus, one can choose $B_D^{\max} = L \cdot T \cdot R_C - \delta$ where $\delta \geq 0$ is small.

In addition, it is clear that the decoder buffer will never underflow, provided that

$$0 \leq B_D^J(T) = L \cdot T \cdot R_C - B_E^J(T), \tag{3.31}$$

or, provided that $B_E^J \leq L \cdot T \cdot R_C$. Therefore, if the encoder buffer satisfies $B_E^{\max} = L \cdot T \cdot R_C$ and never overflows, the decoder buffer never underflows. Herein concludes the simplicity of the constant-rate channel: it is possible to ensure that the decoder buffer does not overflow or underflow simply by ensuring that the encoder buffer does not overflow or underflow.

Next, consider how to choose of the decoder delay L and indicate how the delay enables a variable encoder bit rate, even though the channel has a constant rate. The encoder buffer fullness can be written as

$$B_E^j(T)=\sum_{i=1}^{j} S(iT)-j\cdot T\cdot R_C\leq L\cdot T\cdot R_C \tag{3.32}$$

Eq. (3.32) can be rewritten as

$$\sum_{i=1}^{j} S(i\cdot T)\leq (L+j)\cdot T\cdot R_C\,. \tag{3.33}$$

Thus,

$$L\geq \sum_{i=1}^{j} S(iT)/(T\cdot R_C)-j\,,\text{ for all } j\,. \tag{3.34}$$

Inequality (3.34) indicates the trade-off between the necessary decoder and the variability in the number of encoded bits per picture. Because a variable number of bits per picture can provide better image quality, Inequality (3.34) also indicates the trade-off between the allowable decoder delay and the image quality.

Finally, we explain how Inequality (3.34) involves the variability in the number of bits per coded picture can be seen by examining the two extreme cases of variability. First, suppose that all pictures have the same number of bits $S(iT)=T\cdot R_C$. Then, $L\geq 0$, and no decoder delay is necessary. At the other extreme, suppose all the transmitted bits were for first picture then $L\geq S(T)/(T\cdot R_C)-1$. In this case, the decoder must wait until (most of) the data for the first picture have been received.

Therefore, it is shown in this section that the constant-rate channel provides the simplicity of ensuring no decoder buffer overflow or under-flow by monitoring encoder buffer underflow or overflow. In addition, even though the channel has constant rate, with the use of a delay, it is allowed to obtain some variability in the number of bits per encoded picture.

3.3.2 Leaky-Bucket Channel

Imagine a bucket with a small hole in the bottom. No matter at what rate water enters the bucket, the output flow is at a constant rate, R_C, when there is any water in the bucket, and zero when the bucket is empty. Also, once the bucket is full, any additional water entering it spills over the sides and is lost, i.e. it does not appear in the output stream under the hole. Conceptually, the same idea can be used in modeling the channels.

In this section, we will consider the leaky-bucket channel model. It is shown that for the channel whose rate is controlled by a leaky-bucket policing function the conditions on the encoder bit rate are somewhat weaker than those for a constant-rate channel. Therefore, some additional flexibility can be obtained on the encoder bit rate.

When a leaky-bucket policing function is implemented in a network, an imaginary buffer (it can be called the "bucket") is assumed inside the network and a counter is used indicate the fullness of such buffer. The input to the imaginary buffer is $C(iT)$ bits for the i-th picture period. The output rate of the bucket is C_{output} bits per picture period. The bucket size is $b_{\max}$. Hence, the instantaneous bucket fullness is

$$b_j = \max\{0, b_{j-1} + C(jT) - C_{output}\}. \tag{3.35}$$

If the bucket never underflows, b_j can be written as

$$b_j = \sum_{i=1}^{j} C(iT) - j \cdot C_{output}. \tag{3.36}$$

Note that Eq. (3.36) actually provides only a lower bound on the bucket fullness since the actual bucket fullness may be larger if bucket underflow has occurred.

To ensure that the policing function does not cause high-priority packets to be dropped, rate $C(jT)$ must be such that the bucket never overflows, i.e.,

$$b_j \le b_{\max} \text{ for all } j. \tag{3.37}$$

Or

$$C(jT) \le b_{\max} - b_{j-1} + C_{output}. \tag{3.38}$$

Equation (3.38) defines the leaky-bucket constraint on the rate that is input to the network. It is known from Eq. (3.36) that even if the bucket does underflow, the rate can also be upper bounded by

$$C(jT) \le b_{\max} - \sum_{i=1}^{j-1} C(iT) + j \cdot C_{output}. \tag{3.39}$$

Combining inequalities (3.39) and (3.18), which constrains the rate to prevent encoder buffer underflow and overflow, one has a necessary condition on the encoded rate:

$$
\begin{aligned}
S(jT) &\leq b_{\max} + B_E^{\max} - \sum_{i=1}^{j-1} C(iT) + j \cdot C_{output} - B_E^{j-1}(T) \\
&\leq b_{\max} + B_E^{\max} + j \cdot C_{output} - \sum_{i=1}^{j-1} C(iT) - \left(\sum_{i=1}^{j-1} S(iT) - \sum_{i=1}^{j-1} C(iT) \right) \\
&\leq b_{\max} + B_E^{\max} + j \cdot C_{output} - \sum_{i=1}^{j-1} S(iT).
\end{aligned}
\tag{3.40}
$$

Define the size of a virtual encoder buffer $b_{VE}^{\max} = b_{\max} + B_E^{\max}$ and the fullness of the virtual encoder buffer at picture j as $b_{VE}^{j}(T) = \sum_{i=1}^{j-1} S(iT) - j \cdot C_{output}$. Then, from inequality (3.40), one has

$$
S(jT) \leq b_{VE}^{\max} + C_{output} - b_{VE}^{j-1}(T). \tag{3.41}
$$

Therefore, the encoder accumulated output bit rate $S(jT)$ must be constrained by the encoder's rate-control algorithm to ensure that a virtual encoder buffer of size $b_{VE}^{\max}$ does not overflow, assuming a constant output rate of C_{output} bits per picture. Because this constraint is less strict than preventing an actual encoder buffer with the same drain rate but smaller size $B_E^{\max}$ from overflowing or under-flowing, the leaky-bucket channel has a potential advantage over a channel with constant rate.

However, this is not the only constraint. In fact, preventing decoder buffer overflow can impose a stronger constraint. In particular, the right side of the decoder rate constraint inequality (3.22) may actually be more strict than the leaky-bucket rate constraint inequality (3.38). As a result, one may not actually be able to obtain the full flexibility in the encoder bit rate equivalent to using a virtual encoder buffer of a larger size.

Actually, it is possible to reduce the delay at the decoder without sacrificing the flexibility in the encoded bit rate. Theoretically, one can have the same flexibility in the encoded bit rate, that is available with a constant-rate channel and decoder delay LT, when using a leaky-bucket channel with zero delay, provided that $L = b_{VE}^{\max} / C_{output}$ and $b_{VE}^{\max} = B_E^{\max} = B_D^{\max}$. But, one will certainly have to pay for both $b_{VE}^{\max}$ and C_{output}.

3.4 Compression System with Joint Channel and Encoder Rate-Control

Rate control and buffer regulation is an important issue for both VBR and CBR applications. In the case of VBR encoding, the rate controller attempts to achieve optimum quality for a given target rate. In the case of CBR encoding and real-time application, the rate control scheme has to satisfy the low-latency requirement. Both CBR and VBV rate control schemes have to satisfy buffer constraints. In addition, the rate control scheme has to be applicable to a wide variety of sequences and bit rates.

A rate-control mechanism for video encoder is discussed in this section [3-2]. In this mechanism the number of encoded bits for each video picture and the number of bits transmitted across the variable rate channel can be jointly selected. For a variable bit-rate channel, it is necessary that the decoder buffer imposes a constraint on the transmitted bit rate that is different than that imposed by the encoder buffer. This mechanism also provides the flexibility of having channel bit rates that are less than the maximum allowed rate by the channel, which may be desirable when the channel is not constrained solely by its peak rate.

3.4.1 System Description

A system incorporating these concepts is already shown in Fig.3.1. In this figure, a video signal is applied to the video encoder. The video encoder produces an encoded video bit stream that is stored in the encoder buffer before being transmitted to the variable-rate channel. After being transmitted across the variable-rate channel, the video bit stream is stored in the decoder buffer. The bit stream from the decoder buffer is input to the video decoder, which outputs a decompressed video signal. The delay from encoder buffer input to decoder buffer output, exclusive of channel delay, is exactly LT seconds. The value of the delay L is known a priori, as are the encoder and decoder buffer sizes $B_E^{\max}$ and $B_D^{\max}$.

The rate-control algorithm controls the range of compressed bits output from the encoder. The video encoder produces a bit stream that contains $S(iT)$ number of bits in one picture period, which is within the range given by the encoder rate-control algorithm. These bits are input to the encoder buffer and stored until they are transmitted.

The channel rate-control algorithm takes as input the actual number of bits output in each picture period by the video encoder. It computes estimated accumulated channel rates $C(jT),......,C((j+L-1)\cdot T)$, describing the number of bits that be transmitted across the channel in the following L picture periods. These rates are chosen to prevent encoder and decoder buffer overflow and underflow and to conform to the channel constraint. The channel rate control algorithm sends the estimated value of $C(jT)$ to channel as $\hat{C}(jT)$. If the request is not granted, the channel rate-control algorithm can selectively discard information from the bit stream. However, such information discarding is an emergency measure only since our express purpose is to avoid such discarding. Assume here that the channel grants the request, in which case $C(jT)=\hat{C}(jT)$. If the encoder buffer empties, the transmission is immediately terminated. In most cases, this will cause a reduction of $C(jT)$.

The encoder rate-control algorithm computes a bound on the number of bits that the video encoder may produce without overflowing or under-flowing either the encoder or decoder buffers. It takes as input the actual number of bits $S(jT)$ output in each picture period by the encoder. It also takes as input the channel rate values that are selected by the channel rate-control algorithm. The bound output by the encoder rate-control algorithm is computed to ensure that neither the encoder nor decoder buffers overflow or underflow.

3.4.2 Joint Encoder and Channel Rate-Control Operation

Next, we describe the joint operation of the encoder and channel rate-control algorithms. To simplify the discussion, assume that the channel allows transmission at the requested rate. This is a reasonable assumption since the channel rate-control algorithm is selecting estimated channel rates to conform to the channel constraints negotiated between the channel and the video system.

Joint operation of the encoder and channel rate-control algorithms is described as follows:

1. Initialize buffer fullness variables prior to encoding picture $j=1$; $B_E^j(T)=B_D^j(T)=0$. Also, initialize leaky bucket fullness b_j.
2. Estimate the future channel rates, future leaky-bucket fullness, and future decoder-buffer fullness for the next L pictures. Inequalities (3.22)

and (3.38) are utilized for the channel rates, where for $i \le 0, S(iT) = 0$. Leaky-bucket and decoder-buffer fullness are given by (3.35) and (3.12), respectively. These inequalities can be rewritten as for $i = j, j+1, \ldots\ldots, j+L-1$,

$$S((i-L)\cdot T) - B_D^{i-L-1}(T) \le C(iT) \le S((i-L)\cdot T) + \left(B_D^{\max} - B_D^{i-L-1}(T)\right) \tag{3.42}$$

(The left inequality provides the decoder underflow condition while the right inequality shows the decoder overflow condition.)

$$C(iT) \le b_{\max} - b_{i-1} + C_{output} \tag{3.43}$$

$$b_i = \max\{0, b_{i-1} + C(iT) - C_{output}\} \tag{3.44}$$

$$B_D^{i-L}(T) = B_D^{i-L-1}(T) + C(i\cdot T) - S((i-L)\cdot T). \tag{3.45}$$

Methods for the estimated rates will be discussed in the next section. These methods may ideally consider the fact that a picture with a large number of bits has just occurred or is imminent. They may also consider the cost of transmitting at a given rate. When $j \le L$, no pictures are being decoded and the decoder buffer is only filling. In general, the sum of $C(T), \ldots\ldots, C(LT)$ should be chosen to exceed the expected encoded bit rate of the first few pictures in order to avoid decoder buffer underflow.

1. Compute an upper bound on $C((i+L)\cdot T)$ by using the leaky-bucket constraint (3.43):

$$C((i+L)\cdot T) \le C((i+L)\cdot T)_{upper-bound} = b_{\max} - b_{i+L-1} + C_{output}. \tag{3.46}$$

2. Compute an upper bound on $S(jT)$ using constraints on encoder buffer overflow from inequality (3.16) and decoder buffer underflow from inequality (3.20).

$$S(jT) \le B_E^{\max} + C(jT) - B_E^{j-1}(T) \tag{3.47}$$

$$S(jT) \le C((j+L)\cdot T)_{upper-bound} + B_D^{j-1}(T). \tag{3.48}$$

The minimum of these two upper bounds on $S(jT)$ is output by the encoder rate-control algorithm to the video encoder.

3. Encode picture *j* to achieve $S(jT)$ bits.
4. Using the actual value of $S(jT)$, re-compute $C(jT)$, the actual number of bits transmitted this picture period. (This may be necessary if the encoder buffer would underflow, thus making the actual $C(jT)$ less than that estimated.)

$$C(jT) = B_E^{j-1}(T) + S(jT). \tag{3.49}$$

5. Use obtained $S(jT)$ and $C(jT)$ to compute buffer fullness $B_E^j(T)$, b_j, and $B_D^{j-L}(T)$ by applying Eqs. (3.8), (3.44), and (3.45), respectively.
6. Increment j and go to step 2.

3.4.3 Rate-Control Algorithms

In this section, various encoder rate-control algorithms are introduced and an approach to include the buffer restriction into these algorithms is described. Two channel rate-control algorithms for the leaky bucket are also discussed.

In the encoder rate-control algorithm, the quantizer step size used by the encoder is chosen to ensure not only that the encoder buffer does not overflow and the decoder buffer does not underflow when the corresponding data is decoded, but also that the compressed bit-stream provides best possible quality. In the channel rate-control algorithms, the accumulated channel rate $C(jT)$ is selected based on the channel constraints as well as the decoder buffer fullness.

Encoder Rate-Control Algorithms: To control the encoder rate in order to ensure no encoder or decoder buffer overflow or underflow, one needs to allocate the target bits for each picture and select quantizer step size to achieve the target bits.

Various rate-control methods for bit-allocation and quantizer-step selection are developed in video coding standards, such as MPEG-2 Test Model (TM) [3-14], MPEG-4 Verification Model (VM) [3-15] and H.261 Reference Model (RM) [3-13].

MPEG-2 Rate Control

The MPEG-2 TM rate-control scheme is designed to meet both VBR without delay constraints and CBR with low-latency and buffer constraints. This rate-control algorithm consists of three-steps:

- Target bit allocation: This step estimates the number of bits available to code the next picture. It is performed before coding the picture.
- Rate-control: By means of a "Virtual Buffer", this step sets the reference value of the quantization parameter of each macroblock.
- Adaptive quantization: this step modulates the reference value of the quantization parameter according to the spatial activity in the

macroblock to derive the value of the quantization parameter for providing good visual quality on compressed pictures.

First, the approach to determine bit-allocation for each type of pictures is introduced. Consider the model for bit-allocation method as follows. Assume that the quality of video is measured by using rate-distortion function, e.g. Signal-to-Noise Ratio (SNR) where

$$\text{SNR} \propto \frac{1}{Q},$$

where Q is the average quantization level of the picture. The bit-budget $\hat{S}$ for each picture is based on the linear relation of $1/Q$:

$$\hat{S} \propto \frac{1}{Q}.$$

As one knows, the coded sequence usually consists of three types of pictures, namely, I-, P-, and B-picture. Consider n consecutive pictures (usually a Group Of Pictures (GOP)) with a given coding structure in the video sequence. For example, for the GOP size n=15 with two B-pictures between adjacent P-pictures, the GOP can be IBBPBBPBBPBBPBB. Denote N_I, N_P, N_B to be the number of I-,P-, and B-pictures in the n pictures, respectively. Then,

$$N_I + N_P + N_B = n.$$

Also, denote S_I, S_p, and S_B to be coded bits for the I-, P-, and B-pictures, respectively. Assume that Q_I, Q_p, and Q_B are average quantization level of the I-, P-, and B-pictures, respectively. Define that $X_I = S_I \cdot Q_I$, $X_P = S_P \cdot Q_P$, and $X_B = S_B \cdot Q_B$ are the complexity measures of the I-, P-, and B-pictures, respectively.

If the target video bit-rate is R_b, the goal for the rate-control algorithm is to archive

$$N_P \cdot S_P + N_B \cdot S_B + N_I \cdot S_I = R = \frac{R_b \cdot n}{F_r}, \tag{3.50}$$

with quality balance between different picture types as:

$$\frac{C_I}{Q_I} = \frac{C_P}{Q_P} = \frac{C_B}{Q_B}, \tag{3.51}$$

where $F_r = \frac{1}{T}$ is the picture rate and C_I,C_P, and C_B are constants.

Eq. (3.51) implies

$$Q_P = \frac{C_P}{C_I} Q_I \tag{3.52}$$

and

$$Q_B = \frac{C_B}{C_I} Q_I \tag{3.53}$$

Thus, one can obtain from Eqs. (3.50), (3.51) and (3.53) that

$$\frac{N_I X_I}{Q_I} + \frac{N_P X_P}{\frac{C_P}{C_I} Q_P} + \frac{N_B X_B}{\frac{C_B}{C_I} Q_B} = R, \tag{3.54}$$

Assume that the bit budgets for each picture type satisfy

$$\hat{S}_I = \frac{C_I'}{Q_I},\ \hat{S}_P = \frac{C_P'}{Q_P},\ \text{and}\ \hat{S}_B = \frac{C_B'}{Q_B}.$$

Thus,

$$Q_I = \frac{C_I'}{\hat{S}_I}, Q_P = \frac{C_P'}{\hat{S}_P},\ \text{and}\ Q_B = \frac{C_B'}{\hat{S}_B}. \tag{3.55}$$

From Eqs. (3.54) and (3.55), one has

$$\hat{S}_I \cdot \left(\frac{N_I \cdot X_I}{K_I'} + \frac{N_P \cdot X_P}{K_P'} + \frac{N_B \cdot X_B}{K_B'} \right) = R$$

where K_I', K_P' and K_B' are constants. Therefore,

$$\hat{S}_I = \frac{R}{\frac{N_I \cdot X_I}{K_I'} + \frac{N_P \cdot X_P}{K_P'} + \frac{N_B \cdot X_B}{K_B'}}. \tag{3.56}$$

The bit budgets $\hat{S}_P$ and $\hat{S}_B$ for P- and B-pictures can be derived in a similar manner.

Thus, the bit-allocation and rate-control algorithm can be given as follows:

1. Initialize the number of pictures n and bit budget R. Determine the coding structure with N_I, N_P, and N_B. Initialize or extract from the previous coded pictures: S_I, S_p, and S_B and Q_I, Q_p, and Q_B.
2. If the next picture is I-picture, then compute $\hat{S}_I$. If the next picture is P-picture, then compute $\hat{S}_P$. If the next picture is B-picture, then compute $\hat{S}_B$.
3. For the given picture type, i.e. I- or P- or B-picture, determine the quantization level and code the picture to achieve the bit-budget. If the picture is I-picture, determine Q_I and obtain the coded picture with $S = S_I$ bits and update $X_I = S_I \cdot Q_I$. If the picture is P-picture,

determine Q_P and obtain the coded picture with $S = S_P$ bits and update $X_P = S_P \cdot Q_P$. If the picture is B-picture, determine Q_B and obtain the coded picture with $S = S_B$ bits and update $X_B = S_B \cdot Q_B$.

4. If all n pictures are coded, then Stop; Otherwise compute the bit budget $R = R - S$ for remaining pictures. If the coded picture is I-picture, then set $N_I = N_I - 1$. If the coded picture is P-picture, then set $N_P = N_P - 1$. If the coded picture is B-picture, then set $N_B = N_B - 1$. Go to step 2.

To prevent the encoder buffer either overflow or underflow, the bit-budget $\hat{S}_I$ or $\hat{S}_P$ or $\hat{S}_B$ given in step 2 must be bounded by inequality (3.16). Therefore, the actual implementation of the MPEG-2 TM rate-control algorithm should include a procedure to ensure the condition provided by the inequality (3.16). This process will be discussed further in Chapter 6.

MPEG-4 Rate Control

The rate-control algorithm provided in MPEG-4 Verification Model (VM) is an extension of MPEG-2 TM rate-control algorithm. The main difference is that MPEG-2 TM rate-control algorithm uses a linear rate-distortion model while MPEG-4 VM rate-control algorithm applies a quadratic rate-distortion model.

In MPEG-4 VM rate control, assume that the bit budgets for each picture type satisfy

$$\hat{S}_I = \frac{a_1}{Q_I} + \frac{b_1}{Q_I^2},\ \hat{S}_P = \frac{a_2}{Q_P} + \frac{b_2}{Q_P^2},\ \text{and}\ \hat{S}_B = \frac{a_3}{Q_B} + \frac{b_3}{Q_B^2}. \qquad (3.57)$$

Also, the quality balance between different picture types satisfies

$$K_P \cdot Q_I = K_I \cdot Q_P,\ K_B \cdot Q_P = K_P \cdot Q_B, \qquad (3.58)$$

and

$$\hat{S}_I \cdot N_I + \hat{S}_P \cdot N_P + \hat{S}_B \cdot N_B = R, \qquad (3.59)$$

where K_I, K_P, K_B are constant ratio for I, P, and B pictures, and N_I, N_P, N_B are numbers of pictures to be encoded for I, P, and B pictures, respectively. R represents the remaining number of bits in the current GOP.

Thus, one can solves $\hat{S}_I$, $\hat{S}_P$, and $\hat{S}_B$ from Eqs. (3.57), (3.58), and (3.59). The following steps describe the rate-control algorithm (assuming N_I =1):

1. (Parameter estimation): Collect the bit rate and average quantization step for each type of pictures at the end of encoding each picture. Find the model parameter a_1, a_2, a_3 and b_1, b_2, b_3. The linear regression analysis can be used to find the model parameters [3-15].
2. (Target bit rate calculation): Based on the model found in step 1, one can calculate the target bit rate before encoding. Different formula is used for I, P, and B pictures.
   a. To find $\hat{S}_I$:

$$\begin{aligned}
\alpha &= b_1 + N_P \cdot b_2 \cdot \frac{K_I^2}{K_P^2} + N_B \cdot b_3 \cdot \frac{K_I^2}{K_B^2} \\
\beta &= a_1 + N_P \cdot a_2 \cdot \frac{K_I}{K_P} + N_B \cdot a_3 \cdot \frac{K_I}{K_B} \\
\gamma &= -R \\
\delta &= \beta^2 - 4 \cdot \alpha \cdot \gamma
\end{aligned} \tag{3.60}$$

$$\text{If } \delta < 0, \text{ then } \begin{cases} Q_I^{-1} = -\dfrac{\gamma}{\beta}, \\ \hat{S}_I = a_1 \cdot Q_I^{-1}, \end{cases} \text{ else } \begin{cases} Q_I^{-1} = -\dfrac{\sqrt{\delta} - \beta}{2 \cdot \alpha} \\ \hat{S}_I = a_1 \cdot Q_I^{-1} + a_2 \cdot Q_I^{-2} \end{cases}.$$

   b. To find $\hat{S}_P$:

$$\begin{aligned}
\alpha &= N_P \cdot b_2 + N_B \cdot b_3 \cdot \frac{K_P^2}{K_B^2} \\
\beta &= N_P \cdot a_2 + N_B \cdot a_3 \cdot \frac{K_P}{K_B} \\
\gamma &= -R \\
\delta &= \beta^2 - 4 \cdot \alpha \cdot \gamma
\end{aligned} \tag{3.61}$$

$$\text{If } \delta < 0, \text{ then } \begin{cases} Q_P^{-1} = -\dfrac{\gamma}{\beta}, \\ \hat{S}_P = a_1 \cdot Q_P^{-1}, \end{cases} \text{ else } \begin{cases} Q_P^{-1} = -\dfrac{\sqrt{\delta} - \beta}{2 \cdot \alpha} \\ \hat{S}_P = a_1 \cdot Q_P^{-1} + a_2 \cdot Q_P^{-2} \end{cases}.$$

   c. To find $\hat{S}_B$:

$$\begin{aligned}
\alpha &= N_P \cdot b_2 \cdot \frac{K_B^2}{K_P^2} + N_B \cdot b_3 \\
\beta &= N_P \cdot a_2 \cdot \frac{K_B}{K_P} + N_B \cdot a_3 \\
\gamma &= -R \\
\delta &= \beta^2 - 4 \cdot \alpha \cdot \gamma
\end{aligned} \tag{3.62}$$

$$\text{If } \delta < 0, \text{then} \begin{cases} Q_B^{-1} = -\dfrac{\gamma}{\beta}, \\ \hat{S}_B = a_1 \cdot Q_B^{-1}, \end{cases} \text{else} \begin{cases} Q_B^{-1} = -\dfrac{\sqrt{\delta} - \beta}{2 \cdot \alpha} \\ \hat{S}_B = a_1 \cdot Q_B^{-1} + a_2 \cdot Q_B^{-2} \end{cases}.$$

1. If all n pictures are coded, then Stop; Otherwise update the bit budget R for remaining pictures. If the coded picture is I-picture, then set $N_I = N_I - 1$. If the coded picture is P-picture, then set $N_P = N_P - 1$. If the coded picture is B-picture, then set $N_B = N_B - 1$. Go to step 2.

Again, in order to prevent the encoder buffer either overflow or underflow, the bit-budget $\hat{S}_I$ or $\hat{S}_P$ or $\hat{S}_B$ given in step 2 must be bounded by inequality (3.16). Therefore, the actual implementation of the MPEG-2 TM rate-control algorithm should include a procedure to ensure the condition provided by the inequality (3.16). This process will also be discussed further in Chapter 6.

Both MPEG-2 TM rate-control and MPEG-4 VM rate-control schemes achieve picture level rate control for both VBR and CBR cases. Either a simple linear or a quadratic rate distortion function is assumed in the video encoder. In the case of CBR encoding, a variable picture rate approach is used to achieve the target rate. If a tighter rate control is desired, the same technique is applicable at either slice layer or macroblock layer.

Because of the generality of the assumption, both rate-control schemes are applicable to a variety of bit rates (e.g. 2Mbps to 6Mbps), spatial resolutions (e.g. 720x480 to 1920x1080), temporal resolutions (e.g. 25fps to 30fps), buffer constraints and types of coders (e.g. MC+DCT and wavelet).

H.261 Rate Control

In the H.261 Reference Model encoder [3-13], the quantization level is selected based solely on the fullness of the encoder buffer. With the encoder buffer size $B_E^{\max}$, the buffer control selects

$$Q = 2 \cdot \left\lfloor \frac{32 \cdot B_E^J(T)}{B_E^{\max}} \right\rfloor + 2 \tag{3.63}$$

where $\lfloor \bullet \rfloor$ denotes truncation to a fraction without rounding.

Two simple modifications can be made to this encoder rate-control algorithm. The first modification is introduced to prevent the decoder buffer from under-flowing when the picture currently being encoded is finally decoded. By comparing the constraint of (3.48) to the encoder buffer overflow constraint (3.47), one can set [3-2] :

$$Q = 2 \cdot \left\lfloor 32 \cdot \max\left\{ \frac{B_E^j(T)}{B_E^{\max}}, \frac{B_D^{\max} - B_D^{j+L}(T)}{B_D^{\max}} \right\} \right\rfloor + 2 . \qquad (3.64)$$

Note that the value of the decoder buffer fullness is a prediction of what the decoder buffer fullness is expected to be when the current picture is decoded.

If the channel rate is constant, Eq. (3.64) and Eq. (3.63) give the same quantizer. However, if the channel rate is variable, the quantization control in Eq. (3.64) becomes necessary for preventing the current coded picture from being too large than the system can transmit before this picture being decoded.

However, an additional modification must be made to the quantization strategy to empty the leaky bucket when scene activity is low. If one starts with a full leaky bucket and choose Q as in Eq. (3.64), the leaky bucket would never empty and one would always transmit at the average channel rate. As described in RM [3-13], the quantization level can decrease arbitrarily to increase the number of encoded bits per picture and keep the encoder buffer from under-flowing. However, if one can enable the leaky bucket to empty, the channel rate can subsequently be larger than average, and the leaky-bucket channel can provide better performance than a peak-rate channel. This motivates the second modification on the RM quantization level to obtain some advantages from a variable bit-rate channel.

Rather than encoding fairly static parts of the sequence with progressively smaller quantization levels, the user can use a pre-selected minimum quantization level together with the resultant maximum quality. Therefore, if a scene is less complex, it will be encoded with quantizer $Q = Q_{\min}$, and its average encoded bit rate will be less than C_{output}.

Thus, the quantization level can be chosen as

$$Q = \max\left\{ Q_{\min}, 2 \cdot \left\lfloor 32 \cdot \max\left\{ \frac{B_E^j(T)}{B_E^{\max}}, \frac{B_D^{\max} - B_D^{j+L}(T)}{B_D^{\max}} \right\} \right\rfloor + 2 \right\} . \qquad (3.65)$$

By selecting a minimum quantization level, the user sets an upper bound on the best quality. In general, a given quantization level does not ensure a given image quality. But, the two are closely related. Although the user makes a small quality reduction by choosing $Q_{\min}$, e.g. $Q_{\min} > 4$, such a choice may yield overall better quality.

Leaky-Bucket Channel Rate Control: Two channel rate-control algorithms [3-2] are compared for the leaky bucket. Both use the basic procedure of Section 3.3.2. But, they differ in the selection of $C(j \cdot T)$. The first algorithm is greedy, always choosing the maximum rate allowed by both the channel and the decoder-buffer fullness. The second algorithm is conservative, selecting a channel rate to gradually fill the decoder buffer if the leaky bucker is not full.

Greedy Leaky-Bucket Rate-Control Algorithm (GLB): In this algorithm, the maximum rate is chosen as one both the channel and the decoder buffer will allow. Therefore,

$$C(jT) = \min\left\{B_E^j(T), S((j-L)\cdot T) + B_D^{\max} - B_D^{j-L-1}(T), b_{\max} - b_{j-1} + C_{output}\right\}. \tag{3.66}$$

The first constraint prevents the encoder buffer from under-flowing, the second constraint prevents the decoder buffer from overflowing, and the third constraint prevents the leaky bucket from overflowing. Eq. (3.66) can also be used to estimate the channel rate.

If one only considers the encoder buffer fullness, the GLB algorithm appears optimal. Because data are transmitted at the maximum rate allowed by both the network and the decoder buffer, the encoder buffer is kept as empty as possible, providing the most room to store newly encoded data. If data are transmitted at less than the maximum rate, then the bits remained in the encoder buffer would still need to be transmitted later. However, this algorithm may actually suffer in performance because it fills the bucket as fast as possible. The gain in performance provided by the leaky bucket could be of longer duration if the leaky bucket filled more slowly.

Conservative Leaky-Bucket Rate-Control Algorithm (CLB): The second rate-control algorithm for the leaky bucket is more conservative. The selected rate is the minimum rate among the rate to fill the leaky bucket, the rate to fill the decoder buffer, and the rate to take L pictures to fill the decoder buffer. This estimated rate is computed as

$$C(jT) = \min\left\{r_{j-L}(T), b_{\max} - b_j + C_{output}\right\}, \tag{3.67}$$

where

$$r_{j-L}(T)=\begin{cases} r'_{j-L}(T), & \text{if } r'_{j-L}(T)\le C_{output};\\ C_{output}+\left(r'_{j-L}(T)-C_{putput}\right)/L, & \text{if } r'_{j-L}(T)>C_{output}. \end{cases}$$

$$r'_{j-L}(T)=S((j-L)\cdot T)+B_D^{\max}-B_D^{j-L-1}(T)$$

Because the rate is smaller than the maximum, the duration of the improvement are extended by the leaky bucket, although one may limit the magnitude of the improvement.

Bibliography

[3-1] J. Darragh and R. L. Baker, "Fixed distortion sub-band coding of images for packet-switched networks," IEEE J. Selected Areas Communication, vol.7, no.5, pp. 789-800, June 1989.

[3-2] A. R. Reibman, B. G. Haskell, "Constraints on variable bit-rate video for ATM networks", IEEE Trans. On Circuits and Systems for video technology, Vol. 2, No.4, Dec. 1992.

[3-3] M. Ohanbari, "Two-layer coding of video signals for VBR networks," IEEEJ. Selected Areas Communication., vol.7, no.5, pp.771-781, June 1989.

[3-4] A R. Reibman, "DCT-based embedded coding for packet video," Image Communication, June 1991.

[3-5] G. Karisson and M. Vetterli, "Packet video and its integration into the network architecture," IEEE J. Selected Areas Communication, vol.7, no.5, pp.739-751. June 1989.

[3-6] P. Kithino, K Manabe. Y. Hayahi, and H. Yasuda, "Variable bit-rate coding of video signals for ATM networks," IEEE J. Selected Areas Communication, vol.7, no.5, pp.801-506, June 1989.

[3-7] Naohisa Ohta, Packet Video, Artech House, Inc, Boston, 1994.

[3-8] E. P. Rathgeb, "Modeling and performance comparison of policing mechanisms for ATM networks," IEEE J. Selected Areas Communication, vol.9, no.3, pp. 225-334, April 1991.

[3-9] M. Butto, F. Cavallero and A Tonieti, "Effectiveness of the 'leaky bucket' policing mechanism in ATM networks," IEEE J. Selected Areas Communication., vol.9, no. 3, pp.335-342, April 1991.

[3-10] L. Dittmattn, S. B. Jacobsert, and K. Moth, "Flow enforcement algorithms for ATM networks," IEEE J. Selected Area Communication, vol.9, no.3, pp.343-350, April 1991.

[3-11] Xuemin Chen and Robert O. Eifrig, "Video rate buffer for use with push data flow", US Patent Number 6289129, Assignee: Motorola Inc. and General Instrument Corporation, Sept. 11, 2001.
[3-12] Xuemin Chen, "Rate control for stereoscopic digital video encoding", US Patent Number 6072831, Assignee: General Instrument Corporation, June 6, 2000.
[3-13] "Description of reference models (RM8)." Tech. Rep. 525, CCITT SG-15 Working Party, 1989.
[3-14] Test model editing committee, Test Model 5, MPEG93/457, ISO/IEC JTC1/SC29/WG11, April 1993.
[3-15] Tihao Chiang and Ya-Qin Zhang, "A new rate-control scheme using quadratic rate distortion model", IEEE Transactions on Circuit and Systems for Video Technology, Vol.7, Issue 1, Feb. 1997.
[3-16] Xuemin Chen and Ajay Luthra, "A brief report on core experiment Q2—improved rate control", ISO/IEC JTC1/SC29/WG11, M1422 Maceio, Brazil, Nov. 1996.
[3-17] Xuemin Chen, B. Eifrig and Ajay Luthra, "Rate control for multiple higher resolution VOs: a report on CE Q2", ISO/IEC JTC1/SC29/WG11, M1657, Seville, Spain, Feb. 1997.

4 System Clock Recovery for Video Synchronization

4.1 Video Synchronization Techniques

Digital video systems are unlike analog video systems in two fundamental respects:

- The signal, in its analog state a continuously variable voltage or current, is represented digitally by a limited number of discrete numerical values.
- These numerical values represent the signal only at specific points in time, or sampling instants, rather than continuously at every moment in time.

Sampling instants are determined by various devices. The most common are the analog-to-digital converter (ADC) and the digital-to-analog converter (DAC) that interface between the digital and analog representations of the same signal. These devices will often have a sample clock to control their sampling rate or sampling frequency.

Digital video is often thought to be immune to the many plagues of analog recording and transmission: distortion, various noises, tape hiss, flutter, cross-talk; and if not immune, digital video is certainly highly resistant to most of these maladies. But when practicalities such as oscillator instability, loss connection or noise pickup do intrude, they often affect the digital signal in the time domain as jitter.

Jitter is the variation in the clock signal from nominal. For example, the jitter on a regular clock signal is the difference between the actual pulse transition times of the real clock and the transition times that would have occurred had the clock been ideal, that is to say, perfectly regular.

System jitter occurs as digital video are transmitted through the system, where jitter can be introduced, amplified, accumulated and attenuated, depending on the characteristics of the devices in the signal chain. Jitter in data transmitters and receivers, connection losses, and noise and other spurious signals can all cause jitter and degrade the video signal.

In many digital video applications it is important for the signals to be stored, transmitted, or processed together. This requires that the signals be time-aligned. For example, it is important that the video decoder clock matches the video encoder clock, so that the video signals can be decoded and displayed in the exact time instants. The action of controlling timing in this way is called video (clock) synchronization.

Video synchronization is often required even if the video signals are transmitted through synchronous digital networks because video terminals generally work independently of the network clock. In the case of packet transmission, packet jitter caused by packet multiplexing also has to be considered. This implies that synchronization in packet transmission may become more different than with synchronous digital transmission. Hence, video synchronization functions that consider these conditions should be introduced into video codecs. There are two typical techniques for video synchronization between transmitter and receiving terminals.

One video-synchronization technique measures the buffer fullness at the receiving terminal to control the decoder clock. Fig. 4.1 shows an example of such a technique that uses the digital phase-locked-loop (D-PLL), activated by the buffer fullness. In this technique, a D-PLL controls the decoder clock so that the buffer fullness maintains a certain value. There is no need to insert additional information in the stream to achieve video synchronization.

The other technique requires the insertion of a time reference into the stream at the encoder. At the receiving terminal, the D-PLL controls the decoder clock to keep the time difference between the reference and actual arrival time at a constant value. The block diagram of this technique is shown in Fig. 4.2.

The clock accuracy required for video synchronization will depend on video terminal specifications. For example, a CRT display generally demands an accuracy of less than 10% of a pixel. This means that the required clock stability is about 7.2×10^{-5} for 720 pixels per horizontal video line. It is not difficult to achieve this accuracy if D-PLL techniques are used.

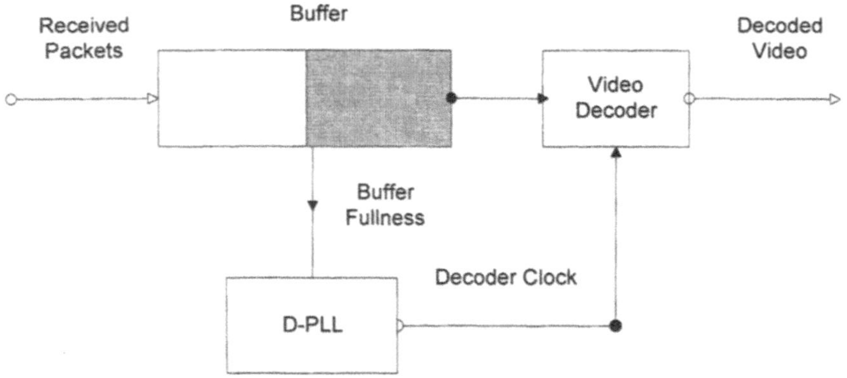

Figure 4.1 Video Synchronization at decoder by using buffer fullness

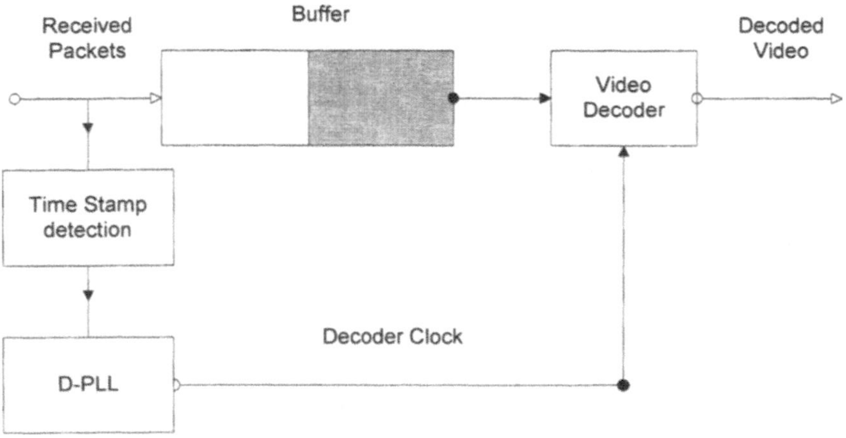

Figure 4.2 Video Synchronization at decoder through time stamping

When a clock is synchronized from an external "sync" source, e.g. timestamps, jitter can be coupled from the sampling jitter of the sync source clock. It can also be introduced in the sync interface. Fortunately, it is possible to filter out sync jitter while maintaining the underlying synchronization. The resulting system imposes the characteristics of a low-pass filter on the jitter, resulting in jitter attenuation above the filter corner frequency.

When sample timing is derived from an external synchronization source in this way, the jitter attenuation properties of the sync systems become important for the quality of the video signal.

In this chapter, we will discuss the technique of video synchronization at decoder through time stamping. As an example, we will focus on MPEG-2 Transport Systems to illustrate the key function blocks of this video synchronization technique. The MPEG-2 system standard [4-1] is widely applied as a transport system to deliver compressed audio and video data and their control signals for various applications such as digital video broadcasting over satellite and cable. The MPEG-2 Systems Layer specifies two mechanisms to multiplex elementary audio, video or private streams to form a program, namely the MPEG-2 Program Stream (PS) and the MPEC-2 Transport Stream (TS) formats. It also provides a function of timing and synchronization of compressed bit streams using time stamps. In error-prone environments such as satellite and cable video networks, the MPEG-2 Transport Stream is the primarily used approach for transporting MPEG-2 streams. As discussed in Chapter 1, an MPEG-2 Transport Stream combines one or more programs into a single fixed-length packet stream. The use of explicit timestamps -- called Program Clock References or PCR in MPEG-2 terminology -- within the packets facilitates the clock recovery at the decoder end ensures synchronization and continuity of MPEG2 Transport Streams. For a brief tutorial of the MPEG-2 Systems Layer the interested reader is referred to [4-1][4-2].

4.2 System Clock Recovery

4.2.1 Requirements on Video System Clock

At the decoder end, application-specific requirements such as accuracy and stability determine the approaches that should be taken to recover the system clock [4-3]. A certain category of applications uses the recovered system clock to directly synthesize a chroma sub-carrier for the composite video signal. The system clock, in this case, is used to derive the chroma sub-carrier, the pixel clock and the picture rate. The composite video sub-carrier must have at least sufficient accuracy and stability so that any normal television receiver's chroma sub-carrier PLL can lock to it, and the chroma signals which are demodulated by using the recovered sub-carrier do not show any visible chrominance phase artifacts. There are often cases in which the application has to meet NTSC, PAL or SECAM specifications for analog televisions [4-4], which are even more stringent. For example, NTSC requires a sub-carrier accuracy of 3 ppm with a maximum long-term drift of 0.1 Hz/sec.

Applications with stringent clock specifications require carefully designed decoders since decoders are responsible of feeding the TV set with a composite signal that meets the requirements. The demodulator in the TV set, as shown in Fig. 4.3, has to extract clock information from this signal for the color sub-carrier regeneration process. The frequency requirements for NTSC specify a tolerance of ±10Hz (or say ±3 ppm) [4-5]. The central sub-carrier frequency is 3.5795454 MHz. The corresponding values for NTSC and PAL composite video are summarized in Table 4.1. The above requirements define the precision of the oscillators for the modulator and thus, the minimum locking range for the PLL at the (decoder) receiver end.

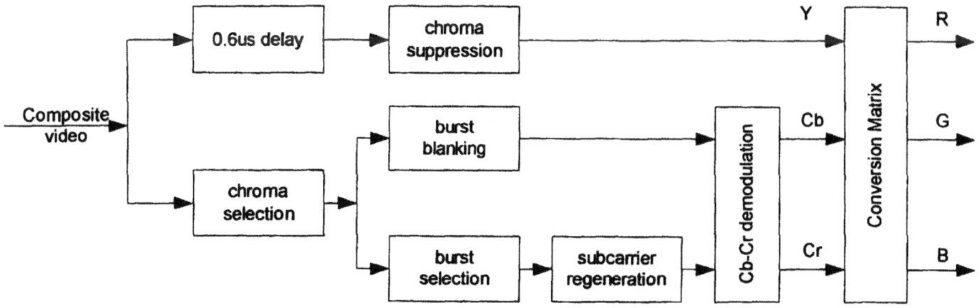

Figure 4.3 A simplified block diagram of PAL/NTSC demodulator system (modified from [4-5]).

There are also requirements for the short- and long-term frequency variations. The maximum allowed short-term frequency variation for an NTSC signal is 56 Hz within a line (or 1 ns/64 ms) whereas the corresponding value for a PAL signal is 69 Hz. This corresponds to a variation of the color frequency of 16 ppm/line in both cases [4-5]. If this requirement is satisfied, a correct color representation can be obtained for each line.

Standard	NTSC	PAL (M)	PAL (B, D, G, H, N)
Frequency (MHz)	3.5795454	3.5756115	4.4336188
Tolerance (Hz)	±10	±10	±5
Tolerance (ppm)	±3	±3	±1

Table 4.1 Requirements on the color sub-carrier for NTSC and PAL (modified from [4-5]).

The maximum long-term frequency variation (clock drift) that the composite NTSC or PAL signal must meet is 0.1 Hz/sec. The drift could be caused by temperature changes at the signal generator and can be determined in an averaging manner over different time-window sizes. In fact, the actual requirement on the color sub-carrier frequency (3.5795454 MHz + 10 Hz for NTSC) in broadcasting applications is an average value that can be measured over any reasonable time period. Averaging intervals in the range from 0.1 second to several seconds are common [4-6].

In MPEG applications, a standard PLL, as shown in Fig. 4.5, is often used to recover the clock from the PCR timestamps transmitted within the stream. The PLL works as follows: Initially, the PLL waits for the reception of the first PCR value for use as the time-base. This value is loaded in the local System Time Clock (STC) counter and the PLL starts operating in a closed-loop fashion. When a new PCR sample is received at the decoder, its value is compared with the value of the local STC. The difference gives an error term. This error term is then sent to a low-pass filter (LPF). The output of the LPF controls the instantaneous frequency of a voltage-controlled oscillator (VCO) whose output provides the decoder's system clock frequency. An analysis on the decoder PLL is given next.

4.2.2 Analysis of the Decoder PLL

The approach of the following analysis is similar to that in [4-7] for traditional PLLs. The main difference here is in the nature of the input signal. The input signal here is assumed to be a linear function as shown in Fig. 4.4, whereas in the case of traditional PLLs, the input signal is usually considered as a sinusoidal function.

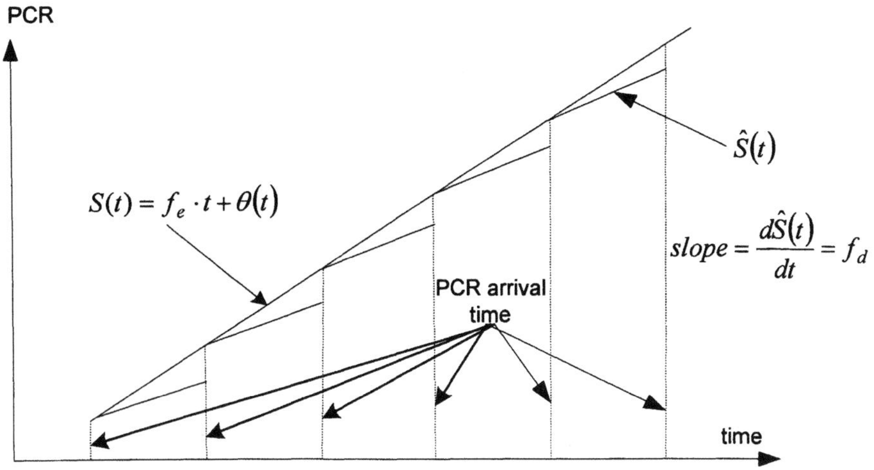

Figure 4.4 Actual PCR function and PCR function used in the analysis (modified from [4-17]).

Although the PCRs arrive at discrete points in time, the incoming PCRs are assumed to form a continuous-time function *s(t)* that is updated at the instants when a new PCR value is received. The incoming clock is modeled with the function

$$S(t) = f_e \cdot t + \theta(t), \tag{4.1}$$

where f_e is the frequency of the encoder system clock and $\theta(t)$ is the incoming clock's phase relative to a designated time origin. As indicated in Fig. 4.4 there is a small discrepancy when modeling the incoming clock signal. The actual incoming clock signal $\hat{S}(t)$ is a function with discontinuities at the time instants at which PCR values are received, with slope equal to f_d for each of its segments, where f_d is the running frequency of the decoder's system clock. For simplicity, however, $S(t)$ is used in place of the actual PCR function $\hat{S}(t)$, since the time between any two consecutive PCR arrivals is bounded by the MPEG-2 standard and equal to at most 0.1 second, which ensures that these two functions are very close.

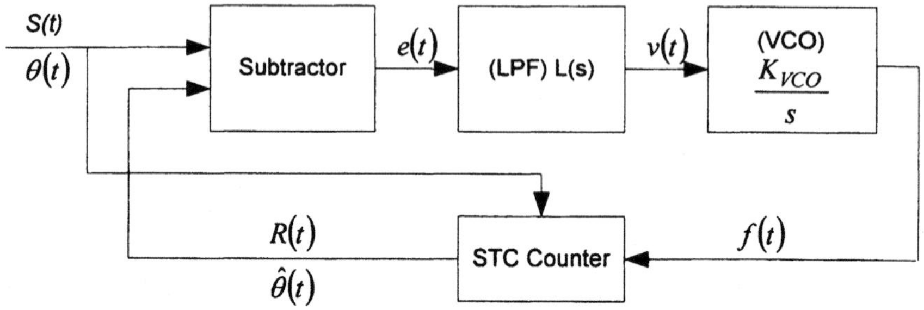

Figure 4.5 A model of the PLL in Laplace-transform domain (modified from [4-17]).

Analogously, the decoder's system time clock (STC) corresponds to the function:

$$R(t) = f_d \cdot t + \hat{\theta}(t), \tag{4.2}$$

where $\hat{\theta}(t)$ is the incoming clock's phase relative to a designated time origin. Therefore, referring to model of the PLL in Fig. 4.5, the error term after the subtractor is given by

$$e(t) = S(t) - R(t) = (f_e - f_d) \cdot t + \left(\theta(t) - \hat{\theta}(t)\right). \tag{4.3}$$

Without loss of generality, assume that $f_e = f_d$. Let us denote this with f_o and move any frequency difference in the phase terms. Now $\theta(t)$ is the input to the control system while $\hat{\theta}(t)$ is the output of the counter as shown in Fig. 4.5. Thus, Eq. (4.3) becomes

$$e(t) = \theta(t) - \hat{\theta}(t) \tag{4.4}$$

The frequency $f(t)$ of the VCO has the nominal frequency f_0 and satisfies $f_0 + K_{VCO} \cdot v(t)$ where K_{VCO} is the gain factor of the VCO. Thus, one has

$$\frac{dR(t)}{dt} = f_0 + K_{VCO} \cdot v(t). \tag{4.5}$$

By definition, one also has

$$R(t) = f_0 \cdot t + \hat{\theta}(t). \tag{4.6}$$

Hence, combining Eq. (4.5) and (4.6) yields

$$\frac{d\hat{\theta}(t)}{dt} = K_{VCO} \cdot v(t) \tag{4.7}$$

From Eq. (4.4) and (4.7) one obtains

$$\begin{aligned}\frac{de(t)}{dt} &= \frac{d\theta(t)}{dt} - K_{VCO} \cdot v(t) \\ &= \frac{d\theta(t)}{dt} - K_{VCO} \int_0^\infty l(t-u)e(u)du.\end{aligned} \tag{4.8}$$

Assume that the Laplace transformations of $e(t)$ and $\theta(t)$ exist and are denoted by $E(s)$ and $\Theta(s)$ respectively, and $L(s)$ is the low-pass filter's transfer function. Eq. (4.8), when transformed to the Laplace domain, becomes

$$s \cdot E(s) = s \cdot \Theta(s) - K_{VCO} \cdot L(s) \cdot E(s) \tag{4.9}$$

Also, assume that $\hat{\theta}(t)$ has a Laplace transform $\hat{\Theta}(s)$. The transfer function $H(s)$ of the closed-loop can be obtained from Eq. (4.9) as

$$H(s) = \frac{\hat{\Theta}(s)}{\Theta(s)} = \frac{K_{VCO} \cdot L(s)}{s + K_{VCO} \cdot L(s)}. \tag{4.10}$$

Eq. (4.10) can also be derived directly from Fig. 4.5 by using $\hat{\Theta}(s) = \frac{K_{VCO}}{s} \cdot L(s) \cdot E(s)$. Then, the Laplace transform $F(s)$ of the recovered frequency function $f(t)$ is given by

$$\begin{aligned}F(s) &= \left(\Theta(s) - \hat{\Theta}(s)\right) \cdot L(s) \cdot K_{VCO} = (1 - H(s)) \cdot \Theta(s) \cdot L(s) \cdot K_{VCO} \\ &= K_{VCO} \cdot \left(1 - \frac{K_{VCO} \cdot L(s)}{s + K_{VCO} \cdot L(s)}\right) \cdot L(s) \cdot \Theta(s) = P(s) \cdot \Theta(s),\end{aligned} \tag{4.11}$$

where $P(s)$ is given by

$$P(s) = K_{VCO} \cdot \left(1 - \frac{K_{VCO} \cdot L(s)}{s + K_{VCO} \cdot L(s)}\right) \cdot L(s). \tag{4.12}$$

Assume that the transfer function of the (loop) low-pass filter is given by

$$L(s) = \frac{K_{LP}}{K_{LP} + s}. \tag{4.13}$$

Thus, the closed-loop transfer function of PLL is

$$H(s) = \frac{K_{VCO} \cdot K_{LP}}{s^2 + K_{LP} \cdot s + K_{VCO} \cdot K_{LP}}. \tag{4.14}$$

It is clear that this is a 2nd -order system and its performance can be characterized by the parameters λ and ω where λ is defined as the damping ratio and ω is defined as natural un-damped frequency and

$$\lambda = \frac{K_{LP}}{2 \cdot \sqrt{K_{LP} \cdot K_{VCO}}}, \quad \omega = \sqrt{K_{LP} \cdot K_{VCO}}. \tag{4.15}$$

The poles S_0 and S_1 can be solved as

$$S_0 = -\lambda \cdot \omega + j\omega \cdot \sqrt{1-\lambda^2}$$
$$S_1 = -\lambda \cdot \omega - j\omega \cdot \sqrt{1-\lambda^2} \quad (4.16)$$

The following are a list of performance parameters defined based on λ and ω. Derivations of these equations can be found in most of control theory textbooks [4-14].

Damping factor: $\alpha = \lambda \cdot \omega$. (4.17)

Damped frequency: $\beta = \omega \cdot \sqrt{1-\lambda^2}$. (4.18)

Setting time: $t_s = \dfrac{4}{\lambda \cdot \omega}$. (4.19)

Maximum overshoot time: $t_{\max} = \dfrac{\pi}{\omega \cdot \sqrt{1-\lambda^2}}$. (4.20)

Maximum overshoot: $M = 1 + e^{-\pi\lambda / \sqrt{1-\lambda^2}}$ (4.21)

All the derivations so far are in the continuous time domain. These derivations can directly be applied to an analog PLL, but the transport design requirement is to build a digital PLL (D-PLL). Normally, the output responses of a discrete-time control system are also functions of continuous-time variable *t*. Therefore, the goal is to map the system that meets the time-response performance requirements specified by λ and ω to a corresponding 2nd-order model in Z-transform domain.

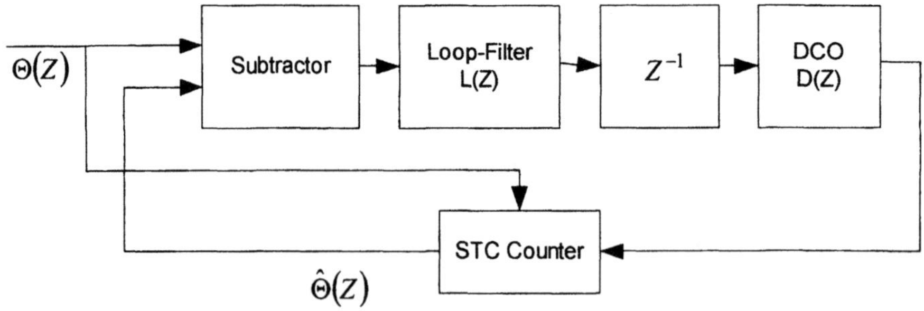

Figure 4.6 A D-PLL model in Z-domain.

A block diagram of the model of a D-PLL is presented in Fig. 4.6.

Transfer functions of each component in the D-PLL are in Z-transfer format as follows:

- The transfer function of the loop filter is

$$L(Z) = \frac{a - Z^{-1}}{1 - Z^{-1}}. \tag{4.22}$$

- The transfer function of a digitally-controlled oscillator (DCO) is

$$D(Z) = \frac{b}{1 - Z^{-1}} \tag{4.23}$$

and Z^{-1} is a delay unit. It is usually a register array.

Based on the block diagram and the above transfer functions, a linear time invariant (LTI) model can be developed to represent the D-PLL with the closed-loop transfer derived as:

$$H(Z) = \frac{a \cdot b \cdot Z - b}{Z^2 + (a \cdot b - 2) \cdot Z + (1 - b)}. \tag{4.24}$$

This is a 2nd -order PLL in Z-domain. By definition of discrete-time transformation, two poles of this system in Z-domain can be mapped from the poles in Laplace transformation domain (Eq. (4.16)) in the following way:

$$Z_0 = e^{S_0 T_s} = e^{-\lambda \cdot \omega \cdot T_s + j\omega T_s \sqrt{1-\lambda^2}},$$

$$Z_1 = e^{S_1 T_s} = e^{-\lambda \cdot \omega \cdot T_s - j\omega T_s \sqrt{1-\lambda^2}}, \tag{4.25}$$

where T_s is the sampling period of the discrete system.

Note that $a \cdot b - 2 = -(Z_1 + Z_2)$ and $1 - b = Z_1 \cdot Z_2$. Thus, with the poles mapped in Z-domain, coefficients *a* and *b* can be derived in a format that is described by parameter Z,

$$a = 2 - 2\frac{e^{-\lambda \cdot \omega \cdot T_s} \cos\left(\omega \cdot T_s \cdot \sqrt{1 - \lambda^2}\right)}{1 - e^{-2\lambda \cdot \omega \cdot T_s}},$$

$$b = 1 - e^{-2\lambda \cdot \omega \cdot T_s}. \tag{4.26}$$

Therefore, if the D-PLL adopts the architecture given by Eq. (4.24), its transfer function will be determined as soon as the poles are mapped.

Usually, the MPEG-2 decoder is synchronized to the source with the PCR stamps by using D-PLL. The decoder keeps a clock reference (STC) and compares it with the PCR stamps. Some "filtering" of the PCR stamps is generally required. If there is a bit error in a PCR, it will cause a temporary spike in the control loop. These spikes should be filtered to prevent unnecessary rate corrections. Over-filtering on PCR can slow the system response to channel changes, or navigation changes.

4.2.3 Implementation of a 2nd-order D-PLL

This section presents detailed information for implementing a completed D-PLL system based on the previous analysis and model mapping results. First of all, a simplified architecture diagram of a 2nd-order D-PLL system is presented in Fig. 4.7.

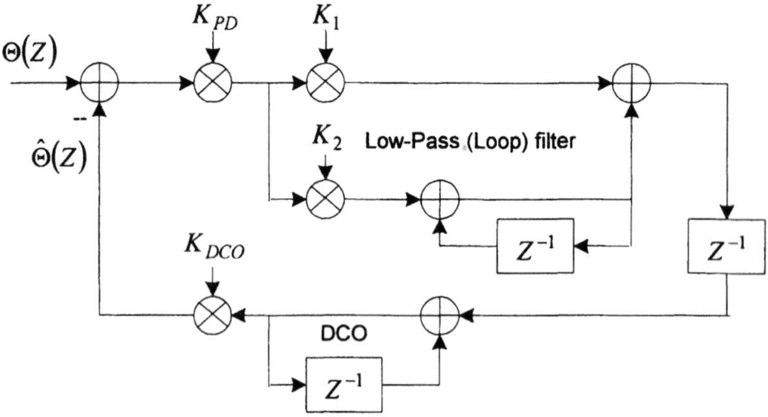

Figure 4.7 A block diagram of a 2nd-order D-PLL system

Based on this architecture, each basic building block is described:

- Low pass (loop) filter: an IIR filter has been designed as the loop-filter, *L(Z)* is its transfer function

$$L(Z) = \frac{K_1 + K_2 - K_1 \cdot Z^{-1}}{1 - Z^{-1}}, \tag{4.27}$$

  where K_1 and K_2 are the gains of the IIR filter.

- A digitally-controlled VCO, or a discrete-time-oscillator, has the transfer function *D(Z)*

$$D(Z) = \frac{K_{DCO}}{1 - Z^{-1}}, \tag{4.28}$$

  where K_{DCO} is the gain of the discrete voltage-controlled-oscillator.

With these building blocks of the D-PLL system, its closed-loop transfer function can be written as:

$$H(Z)=\frac{K_{PD}\cdot L(Z)\cdot D(Z)\cdot Z^{-1}}{1+K_{PD}\cdot L(Z)\cdot D(Z)\cdot Z^{-1}}, \tag{4.29}$$

where, K_{PD} is the gain of the phase detector.

The format of this transfer function can be rewritten as:

$$H(Z)=\frac{(k_1+k_2)\cdot Z-k_1}{Z^2+(k_1+k_2-2)\cdot Z+(1-k_1)}, \tag{4.30}$$

where $k_1=K_{PD}\cdot K_{DCO}\cdot K_1$ and $k_2=K_{PD}\cdot K_{DCO}\cdot K_2$. The denominator of Eq. (4.30) is also called the characteristic equation of the system:

$$\Delta(Z)=Z^2+(k_1+k_2-2)\cdot Z+(1-k_1). \tag{4.31}$$

By using Eq. (4.31), k_1 and k_2 can be resolved based on Eqs. (4.24) and (4.26):

$$k_1=1-e^{-2\lambda\cdot\omega\cdot T_s},$$

$$k_2=1+e^{-2\lambda\cdot\omega\cdot T_s}-2e^{-\lambda\cdot\omega\cdot T_s}\cos\left(\omega\cdot T_s\cdot\sqrt{1-\lambda^2}\right). \tag{4.32}$$

Therefore, with Eq. (4.30) and (4.32), the model of a D-PLL is completely derived.

Stability: One mandatory requirement for designing D-PLLs is that the D-PLL system must be stable. Basically, the stable condition of a discrete-time system is such that the roots of the characteristic equation (4.31) should be inside the unit circle, $|Z| = 1$, in the Z-plane. Normally, after a system is implemented, numerical coefficients can be substituted into the characteristic equation. By solving the characteristic equation numerically, the positions of the poles can be found to determine if the system is stable. However this method is difficult to use to guide the implementation of a D-PLL, since numerical coefficients will not be available at the beginning of the process.

One efficient criterion for testing the stability of a discrete-time system is so called Jury's stability criterion [4-14]. Such criterion can be used to guide designs of a D-PLL to converge to an optimized stable system quickly, without significant amounts of numerical calculation and simulation. It can be directly applied to the 2nd-order D-PLL model to determine the stable condition. According to this criterion, a 2nd-order system with the characteristic equation,

$$\Delta(Z) = A_2 Z^2 + A_1 \cdot Z + A_0, \tag{4.33}$$

should meet following conditions in order to have no roots on, or outside, the unit circle:

$$\begin{aligned} &\Delta(1) > 0, \\ &\Delta(-1) > 0, \\ &|A_0| < A_2. \end{aligned} \tag{4.34}$$

Applying these conditions to Eq. (4.31) the stable conditions for the parameters of this D-PLL architecture are:

$$\begin{aligned} &0 < k_1 < 2, \\ &0 < k_2 < 4. \end{aligned} \tag{4.35}$$

Steady-state errors: A steady-state error analysis of a D-PLL is extremely important in the PLL design. The last paragraph describes the stable conditions of D-PLL system. The steady-state errors of phase and frequency of the D-PLL are studied here. It is proved next that both phase and frequency error of the D-PLL system given by Eq.(4.30) will be zero when the system reaches steady- state.

First consider the phase error. Assume that the phase of the input signal has a step change this can be described by the step function in the time domain:

$$\theta(t) = \Delta\theta \cdot u(t). \tag{4.36}$$

Here $\Delta\theta$ is the constant that the phase of input signal jumped and $u(t) = \begin{cases} 1, & t \geq 0 \\ 0, & t < 0 \end{cases}$. Applying the Z-transform to Eq.(4.36):

$$\Theta(Z) = \frac{\Delta\theta}{1 - Z^{-1}}$$

Based on the linear model given by Eq.(4.30), the output-response function $\Theta(Z)$ of the D-PLL for phase step input can be written as:

$$\hat{\Theta}(Z) = H(Z) \cdot \Theta(Z) = \frac{((k_1 + k_2) \cdot Z - k_1) \cdot \Delta\theta \cdot Z}{(Z - 1) \cdot (Z^2 + (k_1 + k_2 - 2) \cdot Z + (1 - k_1))} \tag{4.37}$$

Based on Eq.(4.37), a numerical analysis can be carried out by using software tools such as MATLAB. Then, the steady-state error of an implemented D-PLL system can be observed. Next, we will focus on some general analysis of this D-PLL system.

First, the phase error is discussed. Assuming *E(Z)* is the phase-error function, by definition, *E(Z)* can be written as follows

$$E(Z) = \Theta(Z) - \hat{\Theta}(Z) = (1 - H(Z)) \cdot \Theta(Z)$$
$$= \frac{\Delta\theta \cdot Z \cdot (Z-1)}{Z^2 + (k_1 + k_2 - 2) \cdot Z + 1 - k_1}. \quad (4.38)$$

According to the Final-Value Theorem,

$$\lim_{n\to\infty} e(nT_s) = \lim_{Z\to 1} (1 - Z^{-1}) \cdot E(Z). \quad (4.39)$$

Based on this theorem, the steady-state error, which is the final value of $e(nT_s)$ in time domain, can be derived. The condition to use the Final-value Theorem is that the function $(1 - Z^{-1}) \cdot E(Z)$ has no poles on or outside the unit circle, $|Z| = 1$, in the z-plane. By substituting Eq.(4.38) into Eq(4.39), one has

$$\lim_{n\to\infty} e(nT_s) = \lim_{Z\to 1} \frac{\Delta\theta \cdot (Z-1)^2}{Z^2 + (k_1 + k_2 - 2) \cdot Z + 1 - k_1} = 0. \quad (4.40)$$

Therefore, one can conclude that when the phase of the input signal had step-jump, the phase error of this D-PLL will eventually be eliminated by the closed-loop system.

Next, the frequency error is considered. For an input signal, assuming $t = 0$, and its frequency jumps from f_1 to f_2, i.e., $\Delta f = f_2 - f_1$. Then, the input phase can be written as follows:

$$\theta(t) = \Delta f \cdot t \cdot u(t). \quad (4.41)$$

By applying a Z-transform to Eq.(4.41), one obtains:

$$\Theta(Z) = \frac{\Delta f \cdot T_s \cdot Z^{-1}}{(1 - Z^{-1})^2}. \quad (4.42)$$

Substituting Eq.(4.42) and Eq.(4.30) into Eq.(4.38), the frequency-error function is derived as:

$$E(Z) = \frac{\Delta f \cdot T_s \cdot Z}{Z^2 + (k_1 + k_2 - 2) \cdot Z + 1 - k_1}. \quad (4.43)$$

Applying the Final-Value Theorem to Eq.(4.43) to get the steady-error in time domain:

$$\lim_{n\to\infty} e(nT_s) = \lim_{Z\to 1} \frac{\Delta\theta \cdot T_s \cdot (Z-1)}{Z^2 + (k_1 + k_2 - 2) \cdot Z + 1 - k_1} = 0. \quad (4.44)$$

Therefore, one can also conclude that when the frequency of input signal has a step jump, the phase error of this D-PLL will eventually be eliminated by the closed-loop system.

4.3 Packetization Jitter and Its Effect on Decoder Clock Recovery

4.3.1 Time-stamping and Packetization Jitter

In jitter-prone environments such as a packet-switched network, the MPEG-2 Transport Stream is also one of approaches for transporting video streams. When transporting MPEG-2 encoded streams over packet-switched networks, several issues must be taken into account. These include the choice of the adaptation layer, method of encapsulation of MPEG-2 packets into network packets, provision of Quality-of-Service (QoS) in the network to ensure control of delay and jitter, and the design of the decoder.

The degradation of the recovered clock at the receiver is introduced primarily by the packet delay variation (jitter). Three different causes contribute to the jitter experienced by an MPEG-2 transport stream as seen at the receiving end: The first is the frequency drift between the transmitter and the receiver clocks, which is usually small compared to the other two causes. The second cause of jitter is due to the packetization at the source, which may displace timestamp values within the stream. Finally, the network may introduce a significant amount of jitter, owing to the variations in queuing delays in the network switches. In this section, our focus is in the second cause, the packetization jitter.

The packetization jitter is mainly caused by the packet encapsulation procedure. In the context of Asynchronous Transfer Mode (ATM) networks, two approaches have been proposed for encapsulation of MPEG-2 Transport Streams in ATM Adaptation Layer 5 (AAL5) packets: the PCR-aware and the PCR-unaware schemes [4-8]. In the PCR-aware scheme, packetization is performed to ensure that a TS packet that contains a PCR is the last packet encapsulated in an AAL-5 packet. This minimizes the PCR jitter during packetization. In the PCR-unaware approach, the sender performs the encapsulation without checking if a PCR is contained in the TS packet. Therefore, the encapsulation procedure could introduce significant jitter to the PCR values. In this case, the presence of jitter introduced by the adaptation layer, may distort the reconstructed clock at the MPEG-2 audio/video decoder. This, in turn, may degrade the quality when the synchronization signals for display of the video frames on the TV set are generated from the recovered clock.

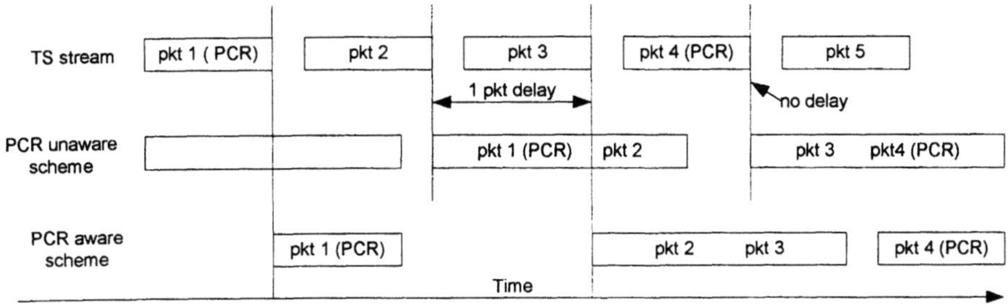

Figure 4.8 Two PCR packing schemes for AAL5 in ATM networks (after [4-17]).

The two schemes are illustrated in Fig. 4.8 [4-17]. In the PCR-unaware case, the packetization procedure does not examine the incoming transport packets and therefore, the second AAL5 Protocol Data Unit (PDU) is the result of encapsulating transport packets 1 and 2, whereas the third AAL5 FDU results from the transport packets numbered 3 and 4. The PCR value in the second AAL5 PDU suffers a delay of one transport packet since it has to wait for the second transport packet to arrive before the PDU is formed. However, this is not the case for the third AAL5 PDU since the PDU becomes complete after the transport packet 4 arrives. On the other hand, the PCR-aware scheme completes a PDU if the current transport packet carries a PCR value. Thus, the second PDU is immediately formed as a result of transport packet 1 which carries a PCR value. The third PDU does not contain any PCR values since it carries transport packets 2 and 3. Finally, the fourth PDU is formed and completed by transport packet 4 in its payload without waiting to receive transport packet 5. It is evident that, for the PCR-unaware case, the process that inserts the PCR values into the MPEG-2 stream at the sender may introduce significant correlation on the resulting jitter of the outgoing transport packets containing PCR values. The PCR-unaware scheme is the recommended method of AAL encapsulation in the ATM Forum Video on Demand specification [4-8].

Several approaches have been reported in the literatures [4-10][4-11][4-12] for the design of the MPEG-2 decoder to reduce the effects of jitter and provide acceptable quality for the decoded video program. The impact of the time-stamping process on the clock recovery at the decoder was extensively studied. The time-stamping process for transporting MPEG-2 over AAL5 using the PCR-unaware packing scheme was reported by Akyildiz et al. [4-8]. It was shown in [4-8] that TS packets containing PCR values may switch indices (between odd and even) in a deterministic manner with a period that depends on both the transport rate and the timer period. This behavior was referred to as "pattern switch." This effect can be avoided by forcing all the PCR values to occupy the same phase in the encapsulated packet stream, or by compensating for the phase difference at the receiver.

In the following sections, several strategies are discussed for performing PCR time-stamping of the MPEG-2 TS. The effects of these strategies on the clock recovery process of the MPEG-2 Systems decoder are analyzed for applications with stringent clock requirements. When the time-stamping scheme is based on a timer with a fixed period, the PCR values in the stream may switch polarity deterministically, at a frequency determined by the timer period and the transport rate of the MPEG signal. This, in turn, can degrade the quality of the recovered clock at the receiver beyond acceptable limits. Three time-stamping methods for solving this problem are considered: (1) selecting the deterministic timer period to avoid the phase difference in PCR values altogether, (2) fine tuning the deterministic timer period to maximize the frequency of PCR polarity changes, and (3) selecting the timer period randomly to eliminate the deterministic PCR polarity changes. For the case of deterministic timer period, the frequency of the PCR polarity changes are derived as a function of the timer period and the transport rate, and use it to find ranges of the timer period for acceptable quality of the recovered clock. A random time-stamping procedure is also discussed based on a random telegraph process [4-13] and lower bounds on the rate of PCR polarity changes are derived such that the recovered clock does not violate the video clock specifications (e.g. PAL and NTSC video).

4.3.2 Possible Input Processes due to PCR-Unaware Scheme

The effects of packetization jitter on the MPEG-2 decoder PLL are analyzed in this subsection. First, let us characterize the input signal at the PLL resulting from the time-stamping and encapsulation schemes at the transmitter. Consider two distinct time-stamping schemes. In the first scheme timestamps are generated by a timer with a deterministic period while, in the second scheme, the timer periods are drawn from a random distribution. In the first case, the pattern switch frequency can be derived as a function of the timer period and transport rate of the MPEG-2 stream, which provides the phase of the input signal at the receiver PLL. In the second case, a random telegraph process is used to model the effect of the time-stamping process, and such process is also used to derive the variance of the recovered clock. This enables us to derive a lower bound on the required rate of change of PCR polarity in the packet stream to maintain the receiver PLL jitter within the specifications.

Due to the time-stamping procedure at the source and the PCR-unaware encapsulation scheme, some effects are resulted on the clock recovery process at the decoder. Since only the tracking performance of the PLL is interested in the discussion, the PLL is assumed to be locked before the input process is applied as the input function $\theta(t)$ of the PLL.

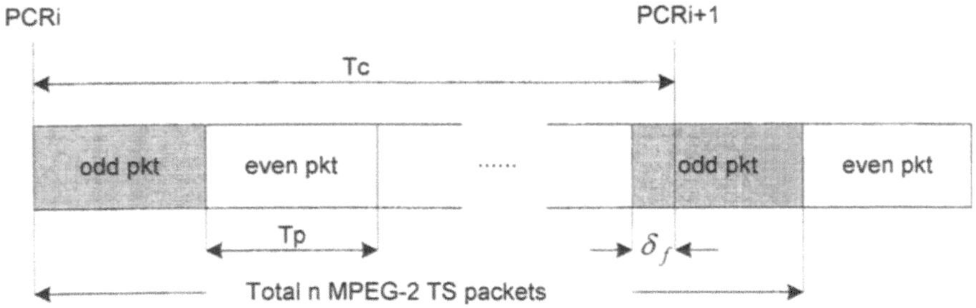

Figure 4.9 Illustration of forward drift of PCR values' packet boundaries when $0 \le \tau < T_p/2$ (after [4-17]).

Under the PCR-unaware scheme, an AAL packet containing two MPEG-2 TS packets may carry a PCR either in the first or in the second TS packet. Therefore, a PCR can suffer one transport packet delay at the destination. Consider the model given in Figure 4.5. Assuming that the PLL is locked before the input process is applied, the resulting phase difference values at its input $\theta(t)$ will be approximately $\pm f_0/2r$, where f_0 is the central frequency in MPEG-2 Systems layer and r is the rate of the MPEG-2 transport stream in packets/second.

First, consider a deterministic case in which a timer with a fixed-period is used to perform the time-stamping procedure.

Deterministic Case: When a timer with a constant period is used at the source to timestamp the MPEG-2 packet stream, the positions of the PCR values switch between even and odd boundaries in the AAL packets at a constant frequency. This effect was observed by Akyildiz et al. [4-8], who referred it as "pattern switch". In this section, the pattern switch frequency is derived as a function of the timer period and the MPEG-2 transport rate. Such a derivation was reported in [4-17].

Let T_p denote the inter-arrival time of MPEG-2 transport packets, and T_c the period of the timer at the transmitter. Since $T_c > T_p$, T_c can be expressed in terms of T_p, as

$$T_c = 2nT_p + \tau \tag{4.45}$$

where n is a non-negative integer and $0 \le \tau < 2T_p$. Since, in general, T_c is not an exact multiple of T_p, the actual time instants at which the PCR values are inserted into the MPEG-2 Transport Stream will drift relative to packet boundaries. More specifically, three cases need to be considered for different ranges of τ.

Case 1: $0 \le \tau < {T_p}/{2}$. In this case, a forward drift of the resulting packet boundaries of the associated PCR values can be identified as illustrated in Fig. 4.9. Let m denotes the integer number of transport packets included in a T_c period, that is,

$$m = \left\lceil \frac{T_c}{T_p} \right\rceil \tag{4.46}$$

Let δ_f denote the forward drift, derived from Fig. 4.9 as

$$\delta_f = T_c - (m-1) \cdot T_p. \tag{4.47}$$

From Eqs. (4.46) and (4.47) one obtains

$$\delta_f = T_c - (\left\lceil \frac{T_c}{T_p} \right\rceil - 1) \cdot T_p. \tag{4.48}$$

It becomes evident from Eq. (4.48) that the number m_c of continuous PCR packets falling into odd or even positions in the MPEG-2 TS is given by

$$\left\lfloor \frac{T_p}{\delta_f} \right\rfloor \le m_c \le \left\lceil \frac{T_p}{\delta_f} \right\rceil. \tag{4.49}$$

Thus, the polarity (even/odd) of timestamp values in the packet stream exhibits a square wave pattern at the input of MPEG-2 decoder's PLL with a period of $2m_cT_c$ and peak-to-peak amplitude of f_oT_p. Therefore the phase of the input signal at the PLL is given by

$$\theta(t) = f_oT_p \sum_{i=-\infty}^{\infty} \left(u\left(t + \frac{i \cdot m_c \cdot T_c}{2}\right) - u\left(t - \frac{i \cdot m_c \cdot T_c}{2}\right)\right) - \frac{f_o \cdot T_p}{2} \quad (4.50)$$

in which $u(t)$ is the unit-step function, i.e. $u(t) = \begin{cases} 1, & t \geq 0 \\ 0, & t < 0 \end{cases}$. If the frequency of the above input signal becomes less than the bandwidth of the PLL, the output of the PLL will follow the pulse with a resulting degradation of the quality of the recovered clock. If the PLL has a perfect LPF, the period of $\theta(t)$ should be less than $1/B_L$. That is,

$$2m_cT_c < \frac{1}{B_L}. \quad (4.50)$$

Case 2: $T_p/2 \leq \tau < 3T_p/2$. In this case, at most two consecutive PCR values may fall into odd- or even-numbered MPEG-2 transport packets. In the specific case that $\tau = T_p$, the PCR values fall in alternate odd- and even-indexed transport packets producing the maximum frequency of changes in timestamp position in the packet stream. The resulting process has high-frequency components that are filtered by the decoder PLL and are unlikely to affect the quality of the recovered clock.

Case 3: $3T_p/2 \leq \tau < 2T_p$. This case is similar to the first one, except that the drift of the packet boundaries of the PCR values is in the backward direction, as shown in Fig. 4.10. In this case, let δ_b denote the backward drift, derived from Fig. 4.10 as

$$\begin{aligned} \delta_b &= mT_p - T_c \\ &= \left\lceil \frac{T_c}{T_p} \right\rceil \cdot T_p - T_c. \end{aligned} \quad (4.51)$$

Similarly, the number of continuous PCR packets falling into only odd or only even positions in the MPEG-2 TS is bounded by the following inequality

$$\left\lfloor \frac{T_p}{\delta_b} \right\rfloor \le m_c \le \left\lceil \frac{T_p}{\delta_b} \right\rceil \tag{4.52}$$

The resulting phase at the input of the PLL, in this case also, is a square-wave with a period of $2m_cT_c$ and peak-to-peak amplitude of f_oT_p. Therefore the input function at the PLL is the same as Eq.(4.50).

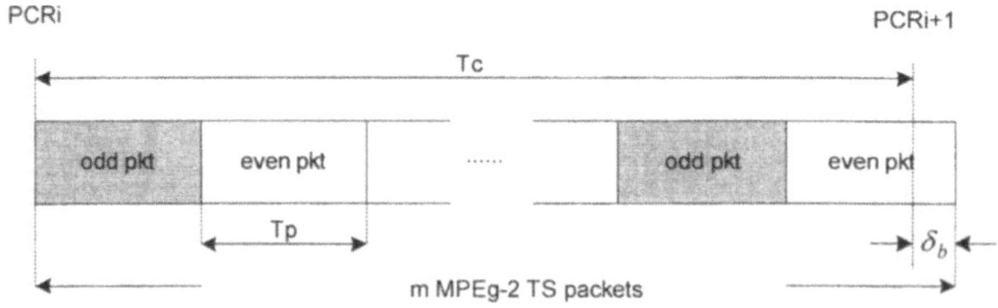

Figure 4.10 Illustration of backward drift of PCR values' packet boundaries when $3T_p/2 \le \tau < 2T_p$ (after [4-17]).

Next, consider a probabilistic case in which the PCR values are placed randomly in the MPEG-2 TS according to a random telegraph process.

Probabilistic Case: MPEG-2 TS with variable inter-PCR delay can be generated by randomizing the time-stamping procedure according to some distribution. In the probabilistic case, assume that the PCR values fall completely in random places in the MPEG-2 Transport Stream. Without loss of generality, also assume that they have the same probability of being in odd or even-indexed transport packets as by using Bernoulli trials. For convenience, such a behavior is analyzed by modeling the input phase $\theta(t)$ as a random telegraph process [4-13].

The objective of the analysis is to obtain the variance or the actual function that describes the recovered clock, i.e., $f(t)$. We derive the variance of the recovered clock in the case that the sequence of $\pm f_0/2r$ values forms a scaled random telegraph process. The random telegraph process $T(t)$ is a random process that assumes values of ± 1, has a mean of zero, and is stationary or cyclo-stationary. Assuming that initially $T(0) = \pm 1$ with equal probability, $T(t)$ is generated by changing polarity with each occurrence of an event of a Poisson process of rate a. In the analysis, a scaled version of the random telegraph process $T(t)$ is used, in which the process gets the values $\pm f_0/2r$. The scaled version is referred by $T_5(t)$. A sample realization of this process is shown in Fig. 4.11.

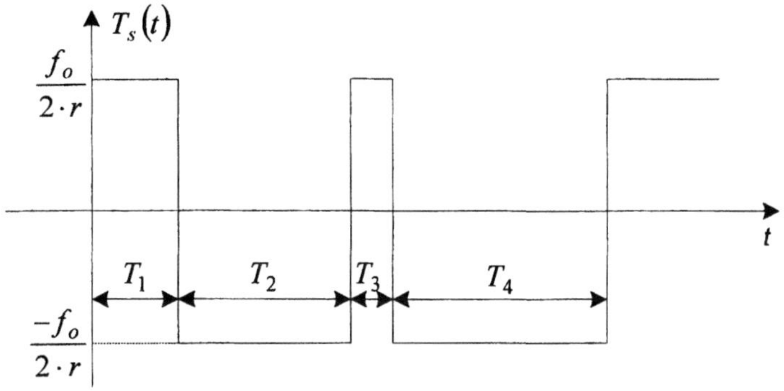

Figure 4.11 Sample realization of $T_S(t)$ process, where T_i are independent identically-distributed (i.i.d.) exponential random variables.

First the statistic measures of the scaled random telegraph process $T_8(t)$ are derived. Since the mean of the random telegraph process is zero, the mean of the scaled version is also zero. Let us now derive the autocorrelation function of $T_8(t)$.

$$
\begin{aligned}
R_{T_s}(t_1, t_2) &= E[T_s(t_1)T_s(t_2)] \\
&= \left(\frac{f_o}{2\cdot r}\right)^2 \text{Prob}[T_s(t_1) = T_s(t_2)] - \left(\frac{f_o}{2\cdot r}\right)^2 \text{prob}[T_s(t_1) \neq T_s(t_2)] \\
&= \left(\frac{f_o}{2\cdot r}\right)^2 \cdot \frac{1 + e^{-2a|t_2 - t_1|}}{2} - \left(\frac{f_o}{2\cdot r}\right)^2 \cdot \frac{1 - e^{-2a|t_2 - t_1|}}{2} \\
&= \left(\frac{f_o}{2\cdot r}\right)^2 \cdot e^{-2a|t_2 - t_1|}.
\end{aligned}
\tag{4.53}
$$

The power spectral density (psd) of the input process is given as the Fourier transform of the auto-correlation function $R_{T_s}(\tau)$. Thus,

$$S_{T_s}(\omega) = \int_{-\infty}^{\infty} R_{T_s}(\tau) e^{-j\omega\tau} d\tau$$
$$= \int_{-\infty}^{\infty} \left(\frac{f_o}{2 \cdot r}\right)^2 \cdot e^{-2a|\tau|} e^{-j\omega\tau} d\tau \tag{4.54}$$
$$= \frac{a \cdot f_o^2}{r^2 \cdot (4a^2 + \omega^2)}.$$

The psd function of the recovered clock is given by

$$S_f(\omega) = S_{T_s}(\omega) |P(\omega)|^2 \tag{4.55}$$

where $|P(w)|$ is the magnitude of the Fourier transform of the function defined in Eq. (4.12). One can obtain the Fourier transform of the signal that has Laplace transform of *P(s)* by substituting S with *jw*. Substituting Eq. (4.13) into Eq.(4.12) yields

$$P(\omega) = \frac{K_{VCO} \cdot K_{LP} \cdot j\omega}{-\omega^2 + K_{LP} \cdot j\omega + K_{VCO} \cdot K_{LP}}. \tag{4.56}$$

From Eqs. (4.55) and (4.56), one has

$$|P(\omega)|^2 = P(\omega) \cdot P^*(\omega) = \frac{(K_{VCO} \cdot K_{LP} \cdot \omega)^2}{(K_{LP} \cdot \omega)^2 + (\omega^2 - K_{VCO} \cdot K_{LP})^2}, \tag{4.57}$$

where $P^*(\omega)$ is the conjugate function of *P(w)*.

The variance of the output process is determined by the inverse Fourier transform of $S_f(\omega)$ at $\tau = 0$. That is,

$$\sigma_{f(t)}^2 = F(S_f(\omega))\Big|_0 = \frac{1}{2\pi} \int_{-\infty}^{\infty} S_f(\omega) \cdot d\omega$$
$$= \frac{1}{2\pi} \int_{-\infty}^{\infty} \left(\frac{a \cdot f_0^2}{r^2 \cdot (4a^2 + \omega^2)}\right) \cdot \left(\frac{(K_{VCO} \cdot K_{LP} \cdot \omega)^2}{(K_{LP} \cdot \omega)^2 + (\omega^2 - K_{VCO} \cdot K_{LP})^2}\right) \cdot d\omega$$
$$= \frac{a \cdot f_o^2 \cdot K_{VCO}^2 \cdot K_{LP}}{r^2 \cdot (8a^2 + 4a \cdot K_{LP} + 2K_{VCO} \cdot K_{LP})}.$$

(4.58)

The above equation provides the variance of the clock at the MPEG-2 Systems decoder. The clock of the color sub-carrier is derived from this clock using a scaling factor that is different for PAL and NTSC. Since the scaled random telegraph process $T_s(t)$ is bounded, one can assume that the recovered clock deviates from its central frequency by at most $\sigma_{f(t)}$. From the requirements for the sub-carrier frequency shown in Table 4.1, the constraints imposed on $\sigma_{f(t)}$ are

$$\sigma_{f_{NTSC}} = \frac{A_{NTSC} \cdot f_o \cdot K_{VCO}}{r} \sqrt{\frac{a \cdot K_{LP}}{8a^2 + 4a \cdot K_{LP} + 2K_{VCO} \cdot K_{LP}}} \leq 10\,\text{Hz}, \tag{4.59}$$

for the recovered NTSC sub-carrier frequency $A_{NTSC} = \frac{3.5795454}{27}$ and

$$\sigma_{f_{PAL}} = \frac{A_{PAL} \cdot f_o \cdot K_{VCO}}{r} \sqrt{\frac{a \cdot K_{LP}}{8a^2 + 4a \cdot K_{LP} + 2K_{VCO} \cdot K_{LP}}} \leq 5\,\text{Hz}, \tag{4.60}$$

for the recovered PAL sub-carrier frequency with $A_{PAL} = \frac{4.43361875}{27}$ (or $A_{PAL} = \frac{3.57561149}{27}$ for PAL-M).

From Inequalities (4.59) and (4.60), two lower bound are obtained on the allowed rate r in packets/second in of NTSC and PAL sub-carriers, so that the clock remains within the specifications.

$$r \geq \frac{A_{NTSC} \cdot f_o \cdot K_{VCO}}{10} \sqrt{\frac{a \cdot K_{LP}}{8a^2 + 4a \cdot K_{LP} + 2K_{VCO} \cdot K_{LP}}} \tag{4.61}$$

for the NTSC case and

$$r \geq \frac{A_{PAL} \cdot f_o \cdot K_{VCO}}{5} \sqrt{\frac{a \cdot K_{LP}}{8a^2 + 4a \cdot K_{LP} + 2K_{VCO} \cdot K_{LP}}} \tag{4.62}$$

for the PAL case.

Analogously, a bound on the rate of change of PCR polarity can be derived so that the clock specification are not violated under a specific transport rate.

$$a \geq \frac{-K_{LP} \cdot \left(4r^2 - C^2\right) + \sqrt{K_{LP}^2 \cdot \left(4r^2 - C^2\right)^2 - 64K_{VCO} \cdot K_{LP} \cdot r^4}}{16r^2},$$

(4.63)

where $C = \frac{A_{NTSC} \cdot f_o \cdot K_{VCO}}{10}$ for the NTSC video and $C = \frac{A_{PAL} \cdot f_o \cdot K_{VCO}}{5}$ for the PAL video.

As an example, inequality (4.61) is applied next to compute the minimum rate for a typical MPEG-2 decoder PLL for the NTSC video. The constant K_{VCO} is used to scale the input signal to the appropriate levels for the MPEG-2 frequency. More specifically, the design of the VCO takes into account the maximum difference in ticks of a 27 MHz clock when the jitter or the PCR inaccuracy due to re-multiplexing operations is at its maximum allowed value, and the limits of the frequency of the decoder. Since, according to MPEG-2 standard [4-3], the maximum jitter expected is around $\pm 0.5ms$, the maximum allowable difference is 13500 ticks. For this maximum difference, the decoder must operate at a frequency within the limits specified in the MPEG-2 standard. That is,

27MHz - 810 Hz $\leq$ decoder clock frequency $\leq$ 27MHz + 810 Hz.

Therefore, the selection of K_{VCO} should be around the value of $\frac{810}{13500}$ or 0.06 in order for the decoder to operate correctly. It is also reasonable to assume that $a \geq 1$ which corresponds to an underlying Poisson process that has a minimum average rate of one arrival every second. Then, the minimum transport rate for the stream to avoid any NTSC clock violations satisfies:

$$r_{NTSC} \geq 0.0214772724 \cdot \sqrt{\frac{a \cdot K_{LP}}{8a^2 + 4a \cdot K_{LP} + 0.12 \cdot K_{LP}}} \quad (4.64)$$

The right side of the inequality is a function of K_{LP} and a. In general, a higher value of a and a lower value of K_{LP} can result a reduced minimum transport rate. A similar result can also be derived for the PAL video.

4.3.3 Solutions for Providing Acceptable Clock Quality

In the previous section, we analyzed and quantified the effect of the time-stamping process at the transmitter on the quality of the recovered clock at the decoder. When the timer-period for time-stamping is chosen deterministically, the pattern switch behavior may manifest itself as a periodic square-wave signal at the input of the decoder PLL for the MPEG-2 transport system. One option to prevent the effect of this pattern switch signal is to eliminate it altogether by forcing all PCR values to occupy the same phase in the AAL packet stream. This would make the receiver clock quality under the PCR-unaware scheme identical to that under the PCR-aware scheme. A second alternative is to maximize the pattern switch frequency by causing the PCR values to switch between odd and even positions in the packet stream at the maximum rate. Finally, a third alternative is to use a random time-stamping interval to avoid the deterministic pattern switch behavior. In this section, the tradeoffs among these approaches are discussed.

The MPEG systems standard [4-3] specifies a maximum interval of 0.1 seconds for transmission of PCR timestamps in the MPEG-2 transport stream. Therefore, in all the schemes that are considered below [4-17], assume that the time-stamping interval T_C is always chosen within this bound.

Scheme 1: Forcing PCR values to stay on one side: The best case in the timestamp process is when the timer period is selected such that the transport rate of the MPEG stream is an exact multiple of the time-stamping rate, that is, the ratio $\frac{T_C}{T_p}$ is an integer. In this case, the PCR values will always fall in either the odd-numbered or the even-numbered transport packets, thus eliminating packetization jitter altogether. Hence, the quality of the recovered clock is similar to that under the PCR-aware case. In practice, however, it is difficult to maintain the time-stamping interval precisely as a multiple of the transport period, because of oscillator tolerances and various quantization effects. These effects may cause the PCR timestamp values to switch polarity at a very low frequency in the packet stream, degrading the quality of the recovered clock over the long term. In addition, loss of packets containing PCR values may cause timestamps to change polarity, that is, an odd-indexed PCR packet may become even-indexed or vice-versa.

Scheme 2: Forcing PCRs to change boundary at high frequency: From the analysis of the previous section, it is clear that the maximum frequency of changes in timestamp position in the packet stream occurs when the time-stamping interval T_C satisfies the equality

$$T_C = (2n+1)\cdot T_p \tag{4.65}$$

where T_p is the transport period of the signal and n is any non-negative integer. If T_C can be chosen precisely to satisfy this inequality, the time-stamped transport packets will occupy alternate (even/odd) positions in the AAL packet stream. The resulting pattern-switch signal is a square wave with the maximum possible frequency among all possible choices of T_C in the range from $2n T_p$ to $2(n+1) T_p$.

Just as in the previous scheme, it is difficult to set T_C precisely to satisfy Eq. (4.65). However, in this case it is not necessary to maintain T_C precisely. In the light of the analysis in the previous section, if the value of the timer period falls in the interval $\lfloor(2n+1/2)\cdot T_p,(2n+3/2)\cdot T_p\rfloor$, the frequency of the resulting pattern-switch pulse is still close to the case when Eq.(4.65) holds. This allows some tolerance for the clocks. Another significant advantage of this scheme is that random losses of packets containing timestamps are unlikely to affect the quality of the reconstructed clock. These hypotheses are verified in many simulation experiments [4-17].

Scheme 3: Random setting of timer period: In this case, the period of the time-stamping timer is set to an arbitrary value, chosen randomly. The same time-stamping interval is chosen for the entire packet stream, resulting in a deterministic pattern-switch signal at the input of the receiver. From the analysis of the previous section, the frequency of the pattern switch signal depends on the relative magnitudes of T_C and T_p. Thus, this scheme needs to be used only when the transport rate of the MPEG signal is not known, since a more intelligent choice can be made when T_C is known.

Scheme 4: Random timer period: Another alternative when the transport rate is not known is to randomize the time-stamping interval, by setting the timer each time to a value drawn from a random distribution. In the previous section, we showed that adequate quality can be maintained for the receiver clock when the time-stamping interval is chosen such that the resulting PCR polarity changes in the packet stream exceeds a minimum rate. Although the analysis was based on modeling the PCR polarity changes with a random telegraph process, in practice similar results can be obtained by choosing the timer period from an exponential distribution. Results from our simulation experiments in the next section indicate that an exponentially-distributed timer period results in almost the same quality for the recovered clock as compared to the case when the PCR polarity changes according to the random telegraph process.

Similar to Scheme 2, this solution does not suffer from degradation of clock quality in the presence of random packet losses. Thus, Scheme 4 is useful when the transport rate T_p is not known with adequate precision.

In summary, Scheme 2 is the preferred scheme when the transport rate of the MPEG signal is known precisely, while Scheme 4 may be used when the transport rate is not known. In the next section, we evaluate the four schemes using both synthetic and real MPEG-2 traces to investigate the characteristics of the recovered clock signal at the receiver under various conditions.

Guidelines for selecting the time-stamping interval are provided for transmission of PCR timestamps in a packetized MPEG-2 transport stream. Based on a systematic analysis of the jitter introduced by the time-stamping process at the receiver, three approaches are identified for setting the timer used to derive the timestamps. In the first approach, the timer period is set precisely so that the transport rate of the MPEG stream is an exact multiple of the time-stamping rate. This completely eliminates packetization jitter, but is difficult to implement in practice because of the precision required in the timer setting. In addition, loss of packets carrying timestamp values can cause the PCR values in the packet stream to switch position, affecting the quality of the recovered clock.

The second approach is to fine-tune the timer period to maximize the frequency of changes in PCR polarity. To maximize the frequency, the time-stamping interval must ideally be set to $T_0/(2n + 1)$, where n is any non-negative integer and T_0 the inverse of the transport rate in packets per second. This causes consecutive PCR values in the packet stream to alternate in polarity. This scheme has the advantage that, even when the timer cannot be set precisely to $T_0/(2n + 1)$, the frequency of PCR polarity changes in the packet stream is still close to ideal. In addition, the scheme is robust in the presence of packet losses. Hence, this is the preferred scheme when the timestamps are generated with a fixed period.

When the transport rate of the MPEG-2 stream is not known and/or when a deterministic timer period is not practical, generating time-stamping intervals randomly (with a certain minimum rate) can still provide adequate quality for the recovered clock. The quality of the decoder clock in this case depends on the process of PCR polarity changes, which, in turn is dependent on the distribution of the time-stamping interval.

Bibliography

For books and articles devoted to video synchronization:

[4-1] A54, Guide to the use of the ATSC digital television standard, Advanced Television Systems Committee, Oct. 19, 1995.
[4-2] Jerry Whitaker, DTV Handbook, 3rd Edition, McGraw-Hill, New York, 2001.
[4-3] ITU-T Recommendation H.222.0 (1995) ISO/IEC 13818-1: 1996, Information technology – Generic coding of moving pictures and associated audio information: Systems.
[4-4] Keith Jack, Video Demystified, HighText Interactive, Inc., San Diego, 1996.
[4-5] G. F Andreotti, G. Michieletto, L. Mon, and A. Profumo, "Clock recovery and reconstruction of PAL pictures for MPEG coded streams transported over ATM networks," IEEE Transactions on Circuits and Systems for Video Technology, vol. 5, pp.508-514, December 1995.
[4-6] D. Fibush. Subearrier, "Frequency, Drift and Jitter in NTSC Systems," ATM Forum, ATM94-0722, July 1994.
[4-7] H. Meyr and G. Ascheid, Synchronization in Digital Communications, John Wiley & Sons, 1990.
[4-8] I. F. Akyiidiz, S. Hrastr, H. Uzunaliogin, and W. Yen, "Comparison and evaluation of packing schemes for MPEG-2 over ATM using AAL5," Proceeding of ICC '96, vol. 3, June 1996.
[4-9] The ATM Forum Technical Committee, Audiovisual Multimedia Services: Video on Demand Specification 1.0, December 1995.
[4-10] P. Hodgins and E. Itakura, "The Issues of Transportation of MPEG over ATM", ATM Forum, ATM94-0570, July 1994.
[4-11] P. Hodgins and E. Itakura, "VBR MPEC-2 over AAL5," ATM Forum, ATM94-1052, December 1994.
[4-12] R. P Singh, Sang-Hoon Lee, and Chong-Kwoon Kim, "Jitter and clock recovery for periodic traffic in broadband packet networks," IEEE Transactions on Communications, vol. 42 No. 5, pp.2189-2196, May 1994.
[4-13] A. Leon Garcia, Probability and Random Processes for Electrical Engineering, Addison-Wesley Publishing Company, second edition, May 1994.
[4-14] Benjamin C. Kuo, Automatic control systems, 7th edition, Prentice-Hall, January 1, 1995.

[4-15] Alan V. Oppenheim and Ronald W. Schafer, Discrete-time signal processing, 2nd edition, Prentice-Hall, Feb. 15, 1999.
[4-16] John L. Stensby, Phase-Locked Loops, Theory and Applications, CRC Press, June 1997.
[4-17] C. Tryfonas, A. Varma, "Time-stamping schemes for MPEG-2 systems layer and their effect on receiver clcok recovery", UCSC-CRL-98-2, University of California, Santa Cruz, 1998.
[4-18] M. De. Prycker, Asynchronous Transfer Mode : Solution for Broadband ISDN, Ellis Horwood, second edition, 1993.
[4-19] S. Dixit and P. Skelly, "MPEG-2 over ATM for video dial tone networks: issues and strategies", IEEE Network, 9(5), pp.30-40 September-October 1995.
[4-20] Y. Kaiser, "Synchronization and de-jittering of a TV decoder in ATM networks", In Proceedings of PV '93, volume 1, 1993.
[4-21] M. Perkins and P. Skelly, "A Hardware MPEG Clock Recovery Experiment in the Presence of ATM Jitter", ATM Forum, May 1994. ATM94-0434.
[4-22] J. Proakis and D. G. Manolakis, Introduction to Digital Signal Processing, Macmillan, 1988.
[4-23] M. Schwartz and D. Beaumont, "Quality of Service Requirements for Audio-Visual Multimedia Services," ATM Forum, July 1994. ATM94-0640.

5 Time-stamping for Decoding and Presentation

5.1 Video Decoding and Presentation Timestamps

As discussed in Chapters 1 and 4, the system clock of a video program is used to create timestamps that indicate the presentation and decoding timing of video, as well as to create timestamps that indicate the instantaneous values of the system clock itself at sampled intervals. The timestamps that indicate the presentation time of video are called Presentation Time Stamps (PTS) while those that indicate the decoding time are called Decoding Time Stamps (DTS). It is the presence of these timestamps and the correct use of the timestamps that provide the facility to synchronize properly the operation of the decoding.

In this chapter, methods for generating the DTS and PTS in the video encoder are discussed. In particular, the time-stamping schemes for MPEG-2 video are introduced as examples. In MPEG-2, a compressed digital video elementary stream is assembled into a packetized elementary stream (PES). Presentation Time Stamps (PTS) are carried in headers of the PES. Decoding Time Stamps (DTS) are also carried in PES headers that have the picture header of an I- or P-picture when bi-directional predictive coding is enabled. The DTS field is never sent with a video PES stream that was generated with B-picture coding disabled. The value for a component of PTS (and DTS, if present) is derived from the 90 kHz portion of the PCR that is assigned to the service to which the component belongs.

Both PTS and DTS are determined in video encoder for coded video pictures. If B-pictures are present in the video stream, coded pictures (sometime also called video access units) do not arrive at the decoder in presentation order. In this case, some decoded pictures in the stream must be stored in a reorder buffer until their correct presentation time (see Fig. 5.1). In particular, I-pictures or P-pictures carried before B-pictures will be delayed in the reorder buffer after being decoded. Any I- or P-picture previously stored in the reorder buffer is presented before the next I- or P-picture is stored. While the I- or P-picture is stored in the reorder buffer, any subsequent B-picture(s) is (are) decoded and presented.

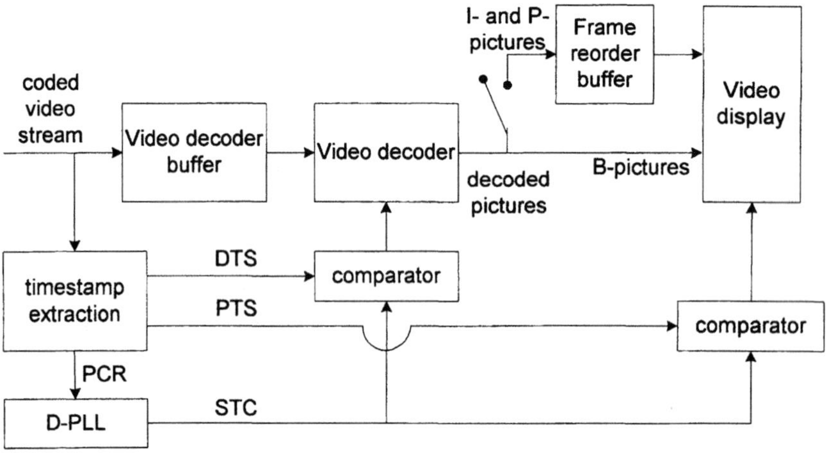

Figure 5.1 A diagram on video decoding by using DTS and PTS.

As shown in Fig. 5.1, the video DTS indicates the time when the associated video picture is to be decoded while the video PTS indicates the time when the presentation unit decoded from the associated video picture is to be presented on the display. Times indicated by PTS and DTS are evaluated with respect to the current system time clock (STC) value. Assume that the decoding time can be ignored. Then, for B-pictures, PTS is always equal to DTS since these pictures are decoded and displayed instantaneously. For I- or P-pictures (if B-pictures are present), PTS and DTS differ by the time that the picture is delayed in the reorder buffer, which will always be a multiple of the nominal picture period, except in film mode. If B-pictures are not present in the video stream, i.e., B-picture type is disabled, all I- and P-pictures arrive in presentation order at the decoder, and consequently their PTS and DTS values are identical. Note that if the PTS and DTS values are identical for a given access unit, only the PTS should be sent in the PES header.

The detailed video coding structures have been reviewed in Chapter 2. The most commonly operated MPEG video coding modes, termed m = 1, m = 2 or m = 3 by the MPEG committee are described as follows. In m = 1 mode, no B-pictures are sent in the coded video stream, and therefore all pictures will arrive at the decoder in presentation order. In m = 2 mode, one B-picture are sent between each I- or P-picture. For example, if pictures arrive at the decoder in the following decoding order:

$I_1\ P_2\ B_3\ P_4\ B_5\ P_6\ B_7\ P_8\ B_9\ I_{10}\ B_{11}$ ......,

they will be reordered in the following presentation order:

$I_1\ B_3\ P_2\ B_5\ P_4\ B_7\ P_6\ B_9\ P_8\ B_{11}\ I_{10}$ ......

In m = 3 mode, two B-picture are sent between each I- or P-picture. Again, for example, if pictures arrive at the decoder in the following decoding order:

$I_1P_2B_3\ B_4\ P_5\ B_6\ B_7\ P_8\ B_9\ B_{10}\ I_{11}B_{12}\ B_{13}\ P_{14}\ B_{15}\ B_{16}$ ......,

They will be reordered in the following presentation order:

$I_1B_3\ B_4\ P_2\ B_6\ B_7\ P_5\ B_9\ B_{10}\ P_8\ B_{12}\ B_{13}\ I_{11}\ B_{15}\ B_{16}\ P_{14}$......

Each time that the picture sync is active, the following picture information are usually required for time stamping of the picture:

- Picture type: I-, P-, or B-picture.
- Temporal Reference: A 10-bit count of pictures in the presentation order.
- Picture Sync Time Stamp (PSTS): A 33-bit value of the 90 kHz portion of the PCR that was latched by the picture sync.

In the normal video mode, the DTS for a given picture is calculated by adding a fixed delay time, T_d, to the PSTS. For some pictures in the film mode, the DTS is generated by (T_d + PSTS - a field time) (this is detailed later in this section). T_d is nominally the delay from the input of the MPEG-2 video encoder to the output of the MPEG-2 video decoder. This delay is also called the end-to-end delay, e.g. $T_d = LT$ for the system discussed in Fig. 3.1 of Chapter 3. In real applications, the exact value of T_d is most likely determined during system integration testing.

The position of the current picture in the presentation order is determined by using the picture type (I, P or B). The number of pictures (if any) for which the current picture is delayed before presentation is used to calculate the PTS from the DTS. If the current picture is a B-picture or if it is an I- or P-picture in m = 1 mode, then it is not delayed in the reorder buffer and the PTS and DTS are identical. In this case, the PTS is sent usually in the PES header that precedes the start of the picture. If the current picture is instead an I- or P-picture and the processing mode is m = 2 or m = 3, then the picture will delayed in the reorder buffer by the total display time required by the subsequent B-picture(s).

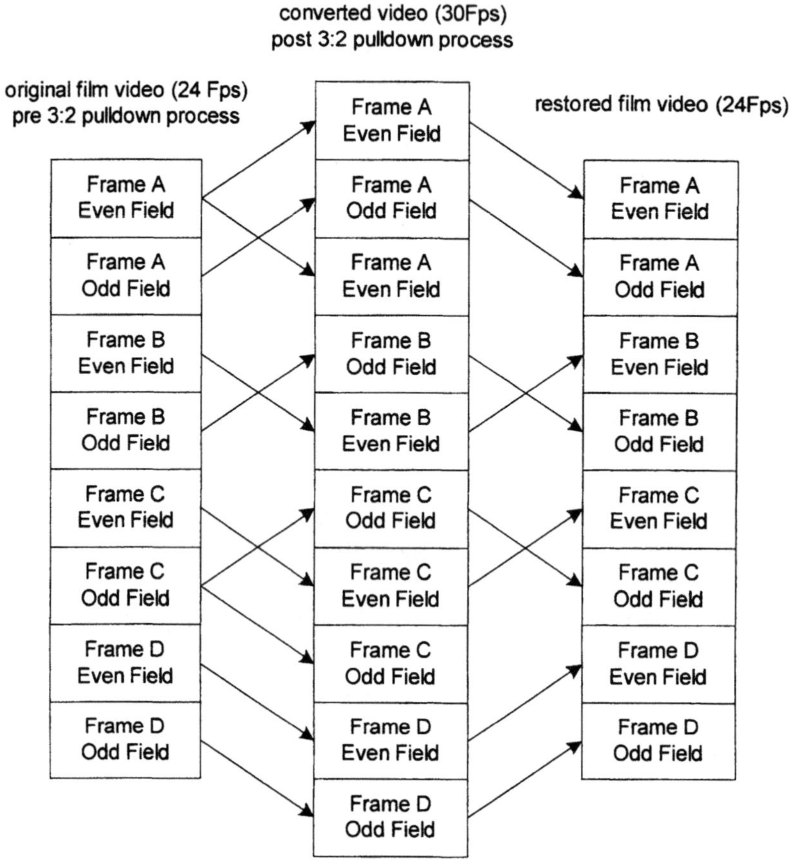

Figure 5.2 Mapping of the 3:2 pull-down process

In addition to considering picture reordering when B-pictures are present, the MPEG-2 video encoder needs to check if the current picture is in the film mode in order to correctly compute the PTS and DTS. In the film mode, two repeated fields have been removed from each ten-field film sequence by the MPEG-2 video encoder, shown in Fig. 5.2. The PSTS will therefore not be stamped on the coded pictures arriving from the MPEG-2 video encoder at the nominal picture rate; two of every four pictures will be of a three-field duration (one and one-half times the nominal picture period), while the other two are of a two-field duration (the nominal picture period). Therefore, in the film mode, the time that an I- or P-picture is delayed in the reordering buffer will not be merely the number of subsequent B-picture(s) multiplied by the nominal picture period, but instead will be the total display time of the B-picture(s). For example, if a P-picture is followed by two B-pictures, one of which will be displayed for a two-field duration and the other for a three-field duration, then the P-picture will be delayed for a total of five-field times, or two and one-half picture times. The PTS for the picture then becomes the DTS plus two and one-half picture times. Note that for NTSC, one-half picture time cannot be expressed in an integral number of 90 kHz clock cycles, and must be either rounded up to 1502 or rounded down to 1501. A fixed set of rules is usually followed for governing rounding of the one-half picture time for the film mode. These rules are outlined later in this chapter.

5.2 Computation of MPEG-2 Video PTS and DTS

In this section, the following commonly used MPEG-2 configurations for the frame-structured picture (i.e. a frame is a picture in this case) are discussed as examples for calculating video PTS and DTS.

- B-picture type disabled (m = 1), non-film mode.
- B-picture type disabled (m = 1), film mode.
- Single B-picture (m= 2), non-film mode.
- Single B-picture (m = 2), film mode.
- Double B-picture (m = 3). non-film mode.
- Double B-picture (m = 3), film mode.

Example 5.1: B-picture Type Disabled, Non-film Mode

In this mode (m = 1), no B-pictures will be sent in the coded video stream. I- and P-pictures in the stream are sent in presentation order, so no picture reordering is required at the MPEG-2 video decoder. Consequently, the PTS and the DTS are identical in this case.

For the i-th picture in the coded video stream, the PTS and DTS are computed as

$$DTS_i = PTS_i = PSTS_i + T_d, \qquad (5.1)$$

where DTS_i = DTS for the i-th picture, PTS_i = PTS for the i-th picture, $PSTS_i$ = $PSTS$ which tags the i-th picture, and T_d = nominal delay from the output of the encoder to the output of the decoder.

If all pictures are processed in non-film mode, then the difference F between PTS_i and PTS_{i-1} should be exactly equal to the nominal picture time in 90 KHz clock cycles (i.e. $F = \frac{90 \times 10^3}{29.97} = 3003$ for NTSC since the picture rate equals 29.97, and $F = \frac{90 \times 10^3}{25} = 3600$ for PAL since the picture rate equals 25).

Therefore, in summary, the following rules can be applied to the calculation of PTS and DTS for the pictures in non-film mode with m = 1.

- Verify to ensure that the difference between PTS_i and PTS_{i-1} is exactly equal to the nominal picture time in 90 KHz clock cycles.
- Calculate PTS and DTS as $DTS_i = PTS_i = PSTS_i + T_d$.
- Send the PTS, but will not send the DTS in the PES header preceding the i-th picture.

Example 5.2: B-picture Type Disabled, Film Mode

Again, I- and P-pictures in a video stream processed without B-pictures (m = 1) are sent in presentation order, regardless of film mode. The PTS and the DTS are therefore identical.

In this case, for the i-th picture in the coded video stream processed in the film mode, the DTS and PTS are calculated by Eq. (5.1) in the same manner as Example 5.1.

Pictures	Transmitted Fields	Repeat First Field Flag	Top Field First Flag	Displayed Field
A	A1	1	1	A1
	A2			A2
				A1
B	B1	0	0	B2
	B2			B1
C	C1	1	0	C2
	C2			C1
				C2
D	D1	0	1	D1
	D2			D2

Table 5.1 Mapping of the 3:2 pull-down flags

Encoded Pictures	Film States	Repeat First Field Flag	Top Field First Flag	Displayed Field
0		0	0	Frm0Bot Frm0Top
1	C	1	0	Frm1Bot Frm1Top Frm1Bot
2	D	0	1	Frm2Top Frm2Bot
3	A	1	1	Frm3Top Frm3Bot Frm3Top
4	B	0	0	Frm4Bot Frm4Top
5	C	1	0	Frm5Bot Frm5Top Frm5Bot
6	D	0	1	Frm6Top Frm6Bot
7	A	1	1	Frm7Top Frm7Bot Frm7Top
8	B	0	0	Frm8Bot Frm8Top
9		0	0	Frm9Bot Frm9Top
10		0	0	Frm10Bot Frm10Top

Table 5.2 A film mode sequence

In the film mode, two flags of MPEG-2 video in the coded picture header, top_field_first and repeat_first_field, are used to indicate the current film-mode state. As shown in Table 5.1 and Fig. 5.2, the four possible film mode states (represented as A, B, C, and D) are repeated in the same order every four pictures. Film mode processing will always commence with state A (or C) and exit with state D (or B). The decoder will display film state A and C pictures for three field times since they both contain a "dropped" field of data. The decoder will re-display the first field to replace the "dropped" field.

This is because in the 3:2 pull-down algorithm, the first field is repeated every other picture to convert film material at 24 pictures/sec to video mode at 30 pictures/second. Film state B and D pictures are displayed for only two field times. A film-mode sequence of four pictures will therefore be displayed as a total of 10 field times. In this way, the decoded video is displayed at the correct video picture rate.

Table 5.2 shows a sequence of eleven coded pictures (m=1) that are output from the video encoder during which the film mode is enabled and then disabled. Picture 0 was not processed in the film mode. Picture 1 is the first picture to be processed in the film mode.

Encoded Pictures	Film States	*PSTSi*	$\Delta PSTS_i$	*DTSi= PTSi*	Display Duration
I_0		$PSTS_0$	F	$PSTS_0+T_d$	F
P_1	C	$PSTS_1$	F	$PSTS_1+T_d$	$\lceil 1.5F \rceil$
P_2	D	$PSTS_2$	$\lceil 1.5F \rceil$	$PSTS_2+T_d$	F
P_3	A	$PSTS_3$	F	$PSTS_3+T_d$	$\lfloor 1.5F \rfloor$
P_4	B	$PSTS_4$	$\lfloor 1.5F \rfloor$	$PSTS_4+T_d$	F
P_5	C	$PSTS_5$	F	$PSTS_5+T_d$	$\lceil 1.5F \rceil$
P_6	D	$PSTS_6$	$\lceil 1.5F \rceil$	$PSTS_6+T_d$	F
I_7	A	$PSTS_7$	F	$PSTS_7+T_d$	$\lfloor 1.5F \rfloor$
P_8	B	$PSTS_8$	$\lfloor 1.5F \rfloor$	$PSTS_8+T_d$	F
P_9		$PSTS_9$	F	$PSTS_9+T_d$	F
P_{10}		$PSTS_{10}$	F	$PSTS_{10}+T_d$	F

Table 5.3. PTS and DTS calculation for the film mode (m = 1)

Unlike the case of non-film and m = 1 mode, the difference between the PSTS tagging successive pictures will not always be equal to the nominal picture time. As can be seen from Table 5.2, the time interval between a picture in film state A and the successive picture in film state B is three field times. Likewise, the time interval between a picture in film state C and the successive picture in film state D is also three field times. Note that for NTSC, three-field time cannot be expressed in an integral number of 90 kHz clock cycles, and must be either rounded up to 4505 or rounded down to 4504. As a convention here, the time interval between the state A picture and the state B picture will always be rounded up to 4505, and the interval between a state C picture and a state D picture will always be rounded down to 4504. Over the four picture film mode sequence, the total time interval will be 4505+3003+4504+3003=15,015 90 KHz clock cycles for NTSC, or exactly five NTSC picture times. Table 5.3 summarizes the PTS and DTS calculations for a sequence of pictures in film mode processed without B-pictures (m = 1).

In summary, the following general rules are applicable to the PTS and DTS for the i-th picture in film mode with m = 1:

- If picture i is in film state C and picture i-1 is in non-film mode, then the difference between $PSTS_i$ and $PSTS_{i-1}$ is F, where F is the nominal picture period in 90 kHz clock cycles (3003 for NTSC, 3600 for PAL).
- If picture i is in film state D, then the difference between $PSTS_i$ and $PSTS_{i-1}$ is $\lceil 1.5F \rceil$, where $\lceil 1.5F \rceil$ is the one and one-half nominal picture periods in 90 kHz clock cycles rounded up to the nearest integer (4505 for NTSC, 5400 for PAL).
- If picture i is in film state A and picture i-1 is in film state D, then the difference between $PSTS_i$ and $PSTS_{i-1}$ is F, where F is the nominal picture period in 90 kHz clock cycles.
- If picture i is in film state B and picture i-1 is in film state A, then the difference between $PSTS_i$ and $PSTS_{i-1}$ is $\lfloor 1.5F \rfloor$, where $\lfloor 1.5F \rfloor$ is the one and on-half nominal picture periods in 90 kHz clock cycles rounded down to the nearest integer (4504 for NTSC, 5400 for PAL).
- If picture i is in non-film mode and picture i-1 is in film state B, then the difference between $PSTS_i$ and $PSTS_{i-1}$ is F, where F is the nominal picture period in 90 kHz clock cycles.
- Compute DTS and PTS as DTSi= PTSi= $PSTS_0+T_d$, where T_d is the nominal delay from the output of the video encoder to the output of the decoder.
- PTS is sent in the PES header preceding the i-th picture.

Example 5.3: Single B-picture, Non-Film Mode

In this mode (m = 2), a single B-picture will be sent between each anchor picture, i.e. each I- or P-picture. If pictures will arrive at the decoder in the following decoding order:

$I_1\ P_2\ B_3\ P_4\ B_5\ P_6\ B_7\ P_8\ B_9\ I_{10}\ B_{11}$ ......,

they will be reordered in the following presentation order:

$I_1\ B_3\ P_2\ B_5\ P_4\ B_7\ P_6\ B_9\ P_8\ B_{11}\ I_{10}$ ......

The MPEG-2 video encoder may generate two types of I-pictures, an I-picture that follows the open Group Of Picture (GOP) or an I-picture that follows the close GOP. An open GOP I-picture will begin a group of pictures to which motion vectors in the previous group of pictures point. For example, a portion of a video sequence is output from the video encoder as

......$P_1B_2P_3B_4I_5B_6P_7B_8$.....

and displayed in Fig. 5.3.

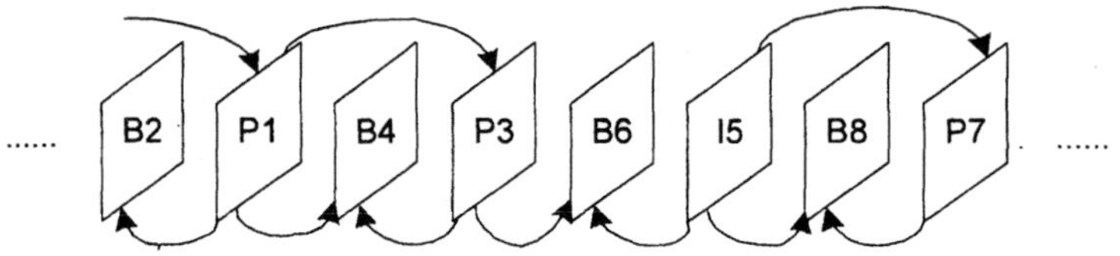

Figure 5.3 An open GOP sequence in display order

The B-picture, B_6, has motion vectors that point to I_5. I_5 is therefore an open GOP I-picture. In m = 2 processing, the video encoder may generate an open GOP I-picture in any position within a video sequence which would normally he occupied by a P-picture.

A closed GOP I-picture will begin a group of pictures that are encoded without predictive vectors from the previous group of pictures. For example, a portion of a video sequence is output from the video encoder as

......$P_1B_2P_3B_4I_5P_6B_7$......

and displayed as in Fig. 5.4.

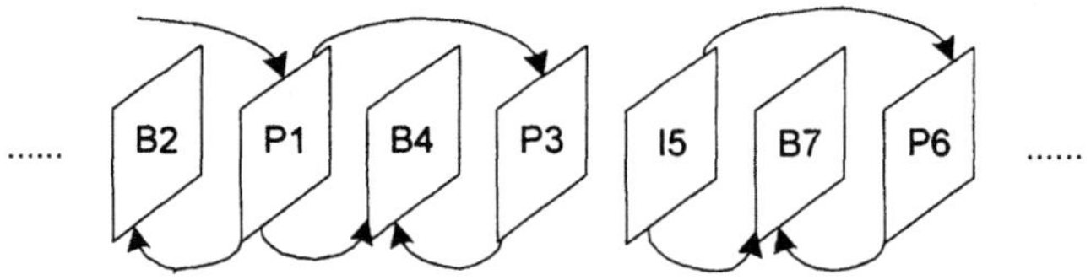

Figure 5.4 A closed GOP sequence in display order

There are no pictures preceding I_5 that contain motion vectors pointing to it. I_5 is therefore a closed GOP I-picture. The video encoder may place a closed GOP I-picture in any point in video sequence.

In MPEG-2 (or MPEG-4 video), the closed GOP (or GOV in MPEG-4) is indicated in the GOP (or GOV) header by the closed_gop (or closed_gov) bit. The picture type and the closed GOP indicator are used to determine the position of the current picture in the display order. The number of pictures (if any) for which the current picture is delayed before presentation is used to calculate the PTS from the DTS as follows:

- If the current picture is a B-picture, then the picture is not delayed in the reorder buffer and the PTS and DTS are identical.
- If the current picture is either an open GOP I-picture or a P-picture that does not immediately precede a closed GOP I-picture, then the picture will be delayed in the reorder buffer by two picture periods while the subsequent B-picture is decoded and displayed. The PTS is equal to the DTS plus twice the picture period.
- If the current picture is either an open GOP I-picture or a P-picture that is immediately before a closed GOP I-picture, then the picture is delayed in the reorder buffer by only one picture period while a previously delayed open I-picture or P-picture is displayed. The PTS is equal to the DTS plus the picture period.
- If the current picture is a closed GOP I-picture, then the picture is delayed one picture period while a previously delayed open GOP I-picture or P-picture is displayed. The PTS is equal to the DTS plus the picture period.

Table 5.4 summarizes the PTS and DTS calculations for a sequence of pictures in non-film mode processed in m = 2 B-picture processing mode. Note that the indices (i) are in decoding order.

Encoded Pictures	Display Order	*PSTSi*	$\Delta PSTS_i$	DTSi	PTSi
I_0 (closed)		$PSTS_0$	*F*	$PSTS_0+T_d$	DTS_0+F
P_1	I_0	$PSTS_1$	*F*	$PSTS_1+T_d$	DTS_1+2F
B_2	B_2	$PSTS_2$	*F*	$PSTS_2+T_d$	DTS_2
I_3 (open)	P_1	$PSTS_3$	*F*	$PSTS_3+T_d$	DTS_3+2F
B_4	B_4	$PSTS_4$	*F*	$PSTS_4+T_d$	DTS_4
P_5	I_3	$PSTS_5$	*F*	$PSTS_5+T_d$	DTS_5+2F
B_6	B_6	$PSTS_6$	*F*	$PSTS_6+T_d$	DTS_6
P_7	P_5	$PSTS_7$	*F*	$PSTS_7+T_d$	DTS_7+F
I_8 (closed)	P_7	$PSTS_8$	*F*	$PSTS_8+T_d$	DTS_8+F
P_9	I_8	$PSTS_9$	*F*	$PSTS_9+T_d$	DTS_9+2F
B_{10}	B_{10}	$PSTS_{10}$	*F*	$PSTS_{10}+T_d$	DTS_{10}

Table 5.4 PTS and DTS calculation for non-film mode (m = 2)

The rules used in computing the DTS and PTS for the i-th picture in non-film mode with m = 2 can be summarized as follows.

- Verify to ensure that the difference between $PSTS_i$ and $PSTS_{i-1}$ is F, where F is the nominal picture period in 90 kHz clock cycles (3003 for NTSC, 3600 for PAL).
- Calculate DTS as $DTS_i = PSTS_i + T_d$, where T_d is the nominal delay from the output of the video encoder to the output of the Decoder.
- If picture i is a B-picture, then $PTS_i = DTS_i$.
- If picture i is a P-picture or an open GOP I-picture and picture i+1 is a closed GOP I-picture, then $PTS_i = DTS_i + F$, where F is the nominal picture period in 90 kHz clock cycles.
- If picture i is a P-picture or an open I-picture and picture i+1 is not a closed GOP I-picture, then $PTS_i = DTS_i + 2F$, where 2F is twice the nominal picture period in 90 kHz clock cycles (6006 for NTSC, 7200 for PAL).
- If picture i is a closed GOP I-picture, then $PTS_i = DTS_i + F$, where F is the nominal picture period in 90 kHz clock cycles.
- If $PTS_i = DTS_i$, then the PTS_i is sent, but the DTS_i will not be sent in the PES header preceding the i-th picture; Otherwise, both the PTS_i and the DTS_i will be sent in the PES header preceding the i-th picture.

Example 5.4. Single B-picture, Film Mode

In the case of m = 2, a sequence of coded pictures will arrive at the decoder in the same picture type order and be likewise reordered in an identical manner regardless of whether film mode was active or inactive when the sequence was coded. The difference between the film mode and non-film mode is the display duration of each picture in the decoder.

As shown in Table 5.2, the display duration of a given picture processed in film mode depends on which of the four possible film mode states (represented as A, B, C, and D) was active when the picture was processed. The video encoder usually needs to implement film mode processing, dropping two fields of redundant information in a sequence of five pictures, prior to predictive coding. Coded pictures with m = 2 in film mode will not output from the video encoder in the A, B, C, D order; the film state order will be rearranged by I-, P-, and B-picture coding. However, after reordering in the decoder, the A, B, C, D order will be re-established prior to display after decoding. The decoder will display film state A and C pictures for three field times since they both contain a "dropped" field of data. Film state B and D pictures are displayed for only two field times.

In m = 2 mode, there are two different scenarios to be examined for developing an algorithm for computing PTS and DTS in film mode. The picture coding type/film state interaction will show two different patterns depending on if the first picture to enter film mode is a B-picture or a P-picture.

Tables 5.5 and 5.6 provide examples of the PTS and DTS calculations for a series of pictures in film mode processed with B-pictures (m = 2). The example shown in Table 5.5 is with a B-picture as the first picture to enter film mode (film state A). Table 5.6 shows the case when a P-picture is the first picture to enter film mode. In here, assume that the encoder doesn't generate a closed GOP I-picture in film mode. The picture coding type/film state pattern in both cases will repeat every fourth picture. Again, the indices (i) are in decoding order.

Encoded Pictures (with film state)	Display Pictures (film with state)	$PSTSi$	$\Delta PSTS_i$	DTSi	PTSi	Display Duration
I_0		$PSTS_0$	F	$PSTS_0+T_d$	DTS_0+F	F
P_1 (B)	I_0	$PSTS_1$	F	$PSTS_1+T_d$	$DTS_1+\lceil 2.5F \rceil$	F
B_2 (A)	B_2 (A)	$PSTS_2$	F	$PSTS_2+T_d$	DTS_2	$\lceil 1.5F \rceil$
P_3 (D)	P_1 (B)	$PSTS_3$	$\lceil 1.5F \rceil$	$PSTS_3+T_d$	$DTS_3+\lfloor 2.5F \rfloor$	F
B_4 (C)	B_4 (C)	$PSTS_4$	F	$PSTS_4+T_d$	DTS_4	$\lfloor 1.5F \rfloor$
I_5 (B)	P_3 (D)	$PSTS_5$	$\lfloor 1.5F \rfloor$	$PSTS_5+T_d$	$DTS_5+\lceil 2.5F \rceil$	F
B_6 (A)	B_6 (A)	$PSTS_6$	F	$PSTS_6+T_d$	DTS_6	$\lceil 1.5F \rceil$
P_7 (D)	I_5 (B)	$PSTS_7$	$\lceil 1.5F \rceil$	$PSTS_7+T_d$	$DTS_7+\lfloor 2.5F \rfloor$	F
B_8 (C)	B_8 (C)	$PSTS_8$	F	$PSTS_8+T_d$	DTS_8	$\lfloor 1.5F \rfloor$
P_9	P_7 (D)	$PSTS_9$	$\lfloor 1.5F \rfloor$	$PSTS_9+T_d$	DTS_9+2F	F
B_{10}	B_{10}	$PSTS_{10}$	F	$PSTS_{10}+T_d$	DTS_{10}	F

Table 5.5. PTS and DTS calculation, film mode, B-picture entry, m =2.

For the case of m=2, the following timing relationships for the i-th picture in film mode need to be satisfied:

- If picture i is in film state A and picture i-1 is in non-film mode, then the difference between $PSTS_i$ and $PSTS_{i-1}$ is F, where F is the nominal picture period in 90 kHz clock cycles (3003 for NTSC, 3600 for PAL).
- If picture i-1 is in film state A, then the difference between $PSTS_i$ and $PSTS_{i-1}$ is $\lceil 1.5F \rceil$, where $\lceil 1.5F \rceil$ is the one and one-half nominal picture

periods in 90 kHz clock cycles rounded up to the nearest integer (4505 for NTSC, 5400 for PAL).

- If picture i-1 is in film state B or D, then the difference between $PSTS_i$ and $PSTS_{i-1}$ is F.
- If picture i-1 is in film stats C, then the difference between $PSTS_i$ and $PSTS_{i-1}$ is $\lfloor 1.5F \rfloor$, where $\lfloor 1.5F \rfloor$ is the one and one-half nominal picture periods in 90 kHz clock cycles rounded down to the nearest integer (4504 for NTSC, 5400 for PAL).

Encoded Pictures (with film state)	Display Pictures (film with state)	$\Delta\, PSTS_j$	DTSi	PTSi	Display Duration
I_0		F	$PSTS_0+T_d$	DTS_0+F	F
P_1 (A)	I_0	F	$PSTS_1+T_d$	DTS_1+2F	F
B_2	B_2	$\lceil 1.5F \rceil$	$PSTS_2+T_d-\lceil 0.5F \rceil$	DTS_2	F
P_3 (C)	P_1 (A)	F	$PSTS_3+T_d-\lfloor 0.5F \rfloor$	$DTS_3+\lceil 2.5F \rceil$	$\lceil 1.5F \rceil$
B_4 (B)	B_4 (B)	$\lfloor 1.5F \rfloor$	$PSTS_4+T_d-\lfloor 0.5F \rfloor$	DTS_4	F
I_5 (A)	P_3 (C)	F	$PSTS_5+T_d-\lceil 0.5F \rceil$	$DTS5+\lfloor 2.5F \rfloor$	$\lfloor 1.5F \rfloor$
B_6 (D)	B_6 (D)	$\lceil 1.5F \rceil$	$PSTS_6+T_d-\lceil 0.5F \rceil$	DTS_6	F
P_7 (C)	I_5 (A)	F	$PSTS_7+T_d-\lfloor 0.5F \rfloor$	$DTS_7+\lceil 2.5F \rceil$	$\lceil 1.5F \rceil$
B_8 (B)	B_8 (B)	$\lfloor 1.5F \rfloor$	$PSTS_8+T_d-\lfloor 0.5F \rfloor$	DTS_8	F
P_9	P_7 (C)	F	$PSTS_9+T_d-\lceil 0.5F \rceil$	$DTS_9+\lfloor 2.5F \rfloor$	$\lfloor 1.5F \rfloor$
B_{10} (D)	B_{10} (D)	F	$PSTS_{10}+T_d$	DTS_{10}	F

Table 5.6. PTS and DTS calculation, film mode, P-picture entry, m =2.

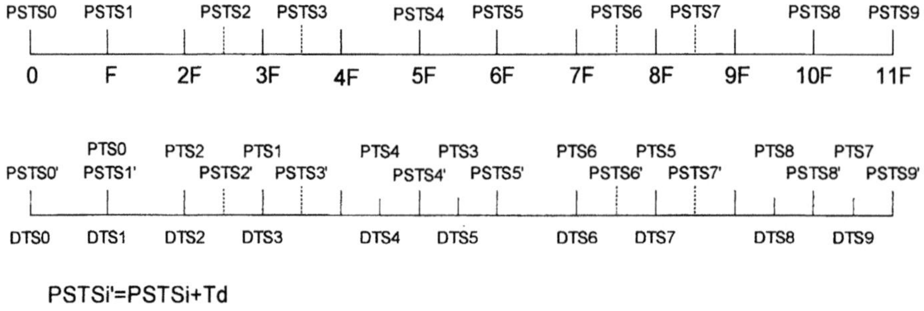

Figure 5.5 Illustration of time stamps relations for the example given in Table 5.6.

The calculation for the PTS and DTS in film mode with m = 2 is conditional to the current picture type and its film state and previous picture's film state. One set of rules for video encoder is summarized in Table 5.7.

Example 5.5: Double B-picture, Non-Film Mode

In m = 3 mode, two B-pictures are sent between each I- or P-picture. As described for the case of non-film, m = 2 mode, the normal I-P- and B-picture order will be altered when a closed GOP I-picture is generated. Again, the picture type and closed GOP indicator are used to determine the position of the current picture in the display order.

Picture Type	Film State for Picture i	Film State for Picture i-1	DTS_i	PTS_i
I (closed)	None	None	$PSTS_i+T_d$	DTS_i+F
B	None	None	$PSTS_i+T_d$	DTS_i
B	A	B		
B	C	D		
B	D	None		
P or I(open)	None	None	$PSTS_i+T_d$	DTS_i+2F
P or I(open)	A	None		
P or I(open)	C	None	$PSTS_i+T_d-\lfloor 0.5F \rfloor$	$DTS_i+\lceil 2.5F \rceil$
P or I(open)	C	D		
P or I(open)	B	None	$PSTS_i+T_d$	$DTS_i+\lceil 2.5F \rceil$
P or I(open)	B	C		
B	None	A	$PSTS_i+T_d-\lceil 0.5F \rceil$	DTSi
B	D	A		
B	B	C	$PSTS_i+T_d-\lfloor 0.5F \rfloor$	DTSi
P or I(open)	A	B	$PSTS_i+T_d-\lceil 0.5F \rceil$	$DTS_i+\lfloor 2.5F \rfloor$
P or I(open)	D	A	$PSTS_i+T_d$	$DTS_i+\lfloor 2.5F \rfloor$

Table 5.7 Calculation of PTS and DTS in the film mode for m=2.

An example of the PTS and DTS calculations is given in Table 5.8 for a coded sequence in non-film mode with m = 3.

The rules used in computing the DTS and PTS for the i-th picture in non-film mode with m = 3 are extension of those rules for m=2. These can be summarized as follows.

- Verify to ensure that the difference between $PSTS_i$ and $PSTS_{i-1}$ is F, where F is the nominal picture period in 90 kHz clock cycles (3003 for NTSC, 3600 for PAL).
- Calculate DTS as $DTS_i = PSTS_i + T_d$, where T_d is the nominal delay from the output of the video encoder to the output of the Decoder.
- If picture i is a B-picture, then $PTS_i = DTS_i$.
- If picture i is a P-picture or an open GOP I-picture and picture i+1 is a closed GOP I-picture, then $PTS_i = DTS_i + F$, where F is the nominal picture period in 90 kHz clock cycles.
- If picture i is a P-picture or an open GOP I-picture and picture i+2 is a closed GOP I-picture, then $PTS_i = DTS_i + 2F$, where 2F is twice the nominal picture period in 90 kHz clock cycles (6006 for NTSC, 7200 for PAL).

Encoded Pictures (i)	Display Order	*PSTSi*	$\Delta PSTS_i$	DTSi	PTSi
I_0 (closed)		$PSTS_0$	*F*	$PSTS_0+T_d$	DTS_0+F
P_3	I_0	$PSTS_1$	*F*	$PSTS_1+T_d$	DTS_1+3F
B_1	B_1	$PSTS_2$	*F*	$PSTS_2+T_d$	DTS_2
B_2	B_2	$PSTS_3$	*F*	$PSTS_3+T_d$	DTS_3
I_6 (open)	P_3	$PSTS_4$	*F*	$PSTS_4+T_d$	DTS_4+3F
B_4	B_4	$PSTS_5$	*F*	$PSTS_5+T_d$	DTS_5
B_5	B_5	$PSTS_6$	*F*	$PSTS_6+T_d$	DTS_6
P_7	I_6	$PSTS_7$	*F*	$PSTS_7+T_d$	DTS_7+F
I_8 (closed)	P_7	$PSTS_8$	*F*	$PSTS_8+T_d$	DTS_8+F
P_{11}	I_8	$PSTS_9$	*F*	$PSTS_9+T_d$	DTS_9+3F
B_9	B_9	$PSTS_{10}$	*F*	$PSTS_{10}+T_d$	DTS_{10}
B_{10}	B_{10}	$PSTS_{11}$	*F*	$PSTS_{11}+T_d$	DTS_{11}
P_{13}	P_{11}	$PSTS_{12}$	*F*	$PSTS_{12}+T_d$	$DTS_{12}+2F$
B_{12}	B_{12}	$PSTS_{13}$	*F*	$PSTS_{13}+T_d$	DTS_{13}
I_{14} (closed)	P_{13}	$PSTS_{14}$	*F*	$PSTS_{14}+T_d$	$DTS_{14}+F$

Table 5.8. An example for the PTS and DTS calculations in non-film mode with m = 3

- If picture i is a P-picture or an open GOP I-picture and pictures i+1 and i+2 are not a closed GOP I-picture, then $PTS_i = DTS_i + 3F$, where 3F is three times the nominal picture period in 90 kHz clock cycles (9009 for NTSC, 10800 for PAL).
- If picture i is a closed GOP I-picture, then $PTS_i = DTS_i + F$, where F is the nominal picture period in 90 kHz clock cycles.
- If $PTS_i = DTS_i$, then the PTS_i is sent, but the DTS_i will not be sent in the PES header preceding the i-th picture; Otherwise, both the PTS_i and the DTS_i will be sent in the PES header preceding the i-th picture.

Example 5.6: Double B-picture, Film Mode

As in the case of m = 2, both the reordering caused by the presence of B-pictures and the difference in display duration for certain film states must be considered when calculating the PTS and DTS for m = 3 pictures in film mode. An example of PTS and DTS calculation for m=3 in film mode is given next.

Encoded Pictures (with film state)	Display Pictures (film with state)	$\Delta PSTS_i$	DTSi	PTSi	Display Duration
I_0		F	$PSTS_0+T_d$	DTS_0+F	F
P_1 (A)	I_0	F	$PSTS_1+T_d$	$DTS_1+\lceil 3.5F \rceil$	F
B_2	B_2 (A)	F	$PSTS_2+T_d$	DTS_2	$\lceil 1.5F \rceil$
B_3	B_3 (B)	$\lceil 1.5F \rceil$	$PSTS_3+T_d$	DTS_3	F
P_4 (D)	P_1 (C)	F	$PSTS_4+T_d$	$DTS_4+\lfloor 3.5F \rfloor$	$\lfloor 1.5F \rfloor$
B_5 (B)	B_5 (D)	$\lfloor 1.5F \rfloor$	$PSTS_5+T_d$	DTS_5	F
B_6 (C)	B_6	F	$PSTS_6+T_d$	DTS_6	F
P_7	P_4	F	$PSTS_7+T_d$	DTS_7+3F	F
B_8	B_8	F	$PSTS_8+T_d$	DTS_8	F
B_9	B_9	F	$PSTS_9+T_d$	DTS_9	F
P_{10}	P_7	F	$PSTS_{10}+T_d$	------	F

Table 5.9 An example for the PTS and DTS calculations in the film mode with m = 3 and entering at the first B-picture.

Encoded Pictures (with film state)	Display Pictures (film with state)	$\Delta PSTS_i$	DTSi	PTSi	Display Duration
I_0		F	$PSTS_0+T_d$	DTS_0+F	F
P_1 (A)	I_0	F	$PSTS_1+T_d$	$DTS_1+\lceil 3.5F \rceil$	F
B_2	B_2 (A)	F	$PSTS_2+T_d$	DTS_2	F
B_3	B_3 (B)	F	$PSTS_3+T_d$	DTS_3	$\lceil 1.5F \rceil$
P_4 (D)	P_1 (C)	$\lceil 1.5F \rceil$	$PSTS_4+T_d$	$DTS_4+\lfloor 3.5F \rfloor$	F
B_5 (B)	B_5 (D)	F	$PSTS_5+T_d$	DTS_5	$\lfloor 1.5F \rfloor$
B_6 (C)	B_6	$\lfloor 1.5F \rfloor$	$PSTS_6+T_d$	DTS_6	F
P_7	P_4	F	$PSTS_7+T_d$	$DTS_7+ 3F$	F
B_8	B_8	F	$PSTS_8+T_d$	DTS_8	F
B_9	B_9	F	$PSTS_9+T_d$	DTS_9	F
P_{10}	P_7	F	$PSTS_{10}+T_d$	------	F

Table 5.10 An example for the PTS and DTS calculations in the film mode with m = 3 and entering at the second B-picture.

Encoded Pictures (with film state)	Display Pictures (film with state)	$\Delta PSTS_i$	DTSi	PTSi	Display Duration
I_0		F	$PSTS_0+T_d$	DTS_0+F	F
P_1 (A)	I_0	F	$PSTS_1+T_d$	DTS_1+3F	F
B_2	B_2	$\lceil 1.5F \rceil$	$PSTS_2+T_d-\lceil 0.5F \rceil$	DTS_2	F
B_3	B_3	F	$PSTS_3+T_d-\lfloor 0.5F \rfloor$	DTS_3	F
P_4 (D)	P_1 (A)	$\lfloor 1.5F \rfloor$	$PSTS_4+T_d-F$	DTS_4+4F	$\lceil 1.5F \rceil$
B_5 (B)	B_5 (B)	F	$PSTS_5+T_d-\lceil 0.5F \rceil$	DTS_5	F
B_6 (C)	B_6 (C)	F	$PSTS_6+T_d-\lfloor 0.5F \rfloor$	DTS_6	$\lfloor 1.5F \rfloor$
P_7	P_4 (D)	F	$PSTS_7+T_d$	$DTS_7+ 3F$	F
B_8	B_8	F	$PSTS_8+T_d$	DTS_8	F
B_9	B_9	F	$PSTS_9+T_d$	DTS_9	F
P_{10}	P_7	F	$PSTS_{10}+T_d$	------	F

Table 5.11 An example for the PTS and DTS calculations in the film mode with m = 3 and entering at P-picture.

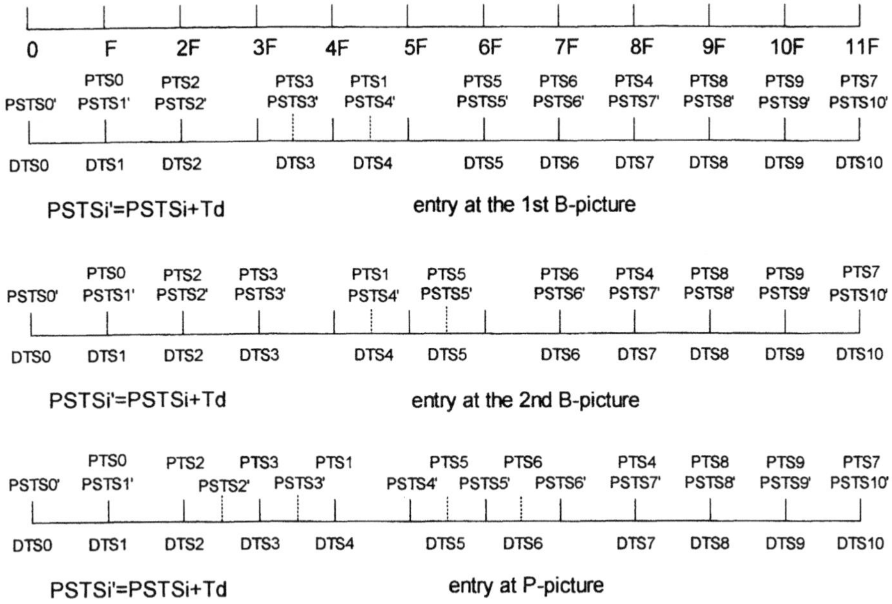

Figure 5.6. Time stamps relations of examples given in Tables 5.9, 5.10 and 5.11.

The general rules for m=3 in the film-mode can also be determined in a similar manner as that for m=2 in the film mode. Interested readers can develop these rules as exercises.

Time Stamp Errors: As discussed in Chapter 4, the clock-recovery process is designed to track the encoder timing and manage the absolute and relative system timing for video and other multimedia data during decoding operations. Specifically, the clock-recovery process monitors timestamps in the transport stream and update the system clock in a multimedia program when necessary.

During MPEG-2 decoding, the clock-recovery process is programmed to monitor the PCRs in the transport stream. The clock-recovery process uses PCRs in the stream against its own system clock, and indicates discontinuities every time an error is seen in a PCR that is larger than a programmable threshold. If a PCR discontinuity is detected in the incoming transport stream, the new PCR is used to update the video system clock counter (STC). After the video decoder STC is updated, the PLL begins to track PCR. The picture is decoded when DTS = STC.

However, the network jitters can cause time-stamp errors that, in turn, could cause decoder buffer over- or under-flows. Therefore, at any moment, if the decoder buffer is overflowing, some coded pictures in the buffer will be dropped without decoding. If DTS = STC, but the decoder buffer is still underflow, the decoder can wait certain amount of time for the current coded picture completely entering the buffer for decoding. In these cases, error-concealment algorithms are usually required.

The above methods of calculating DTS and PTS for MPEG-2 video can be directly used in (or be generalized to) other compressed video such as MPEG-4 video [5-10] and H.263 video [5-12].

Bibliography

[5-1] ITU-T Recommendation H.222.0 (1995) ISO/IEC 13818-1: 1996, Information technology — Generic coding of moving pictures and associated audio information: Systems.
[5-2] Xuemin Chen, "Synchronization of a stereoscopic video sequence", US Patent Number 5886736, Assignee: General Instrument Corporation, March 23, 1999.
[5-3] Xuemin Chen and Robert O. Eifrig, "Video rate buffer for use with push data flow", US Patent Number 6289129, Assignee: Motorola Inc. and General Instrument Corporation, Sept. 11, 2001.
[5-4] WO9966734, Xuemin Chen, Fan Lin, and Ajay Luthra, "Video encoder and encoding method with buffer control", 2000.
[5-5] Xuemin Chen, "Rate control for stereoscopic digital video encoding", US Patent Number 6072831, Assignee: General Instrument Corporation, June 6, 2000.
[5-6] Jerry Whitaker, DTV Handbook, 3rd Edition, McGraw-Hill, New York, 2001.
[5-7] Naohisa Ohta, Packet Video, Artech House, Inc, Boston, 1994.
[5-8] B. G. Haskell, A. Puri, and A. N. Netravali, Digital Video: An Introduction to MPEG-2, New York: Chapman & Hall, 1997.
[5-9] Atul Puri and T. H. Chen, Multimedia Standards and Systems, Chapman & Hall, New York, 1999.

[5-10] ISO/IEC 14496-2:1998, Information Technology - Generic coding of audio-visual objects - Part 2: Visual.
[5-11] Test model editing committee, Test Model 5, MPEG93/457, ISO/IEC JTC1/SC29/WG11, April 1993.
[5-12] ITU-T Experts Group on Very Low Bitrate Visual Telephony, "ITU-T Recommendation H.263: Video Coding for Low Bitrate Communication," Dec. 1995.

6 Video Buffer Management and MPEG Video Buffer Verifier

6.1 Video Buffer Management

The rate-buffer management in video encoder provides a protocol to prevent the decoder buffer under- and/or over-flows. With such a protocol, adaptive quantization is applied in the encoder along with rate-control to ensure the required video quality and to satisfy the buffer regulation. In Chapter 3, we have derived the buffer dynamics and determined general conditions for preventing both encoder and decoder buffers under- and/or over-flow, e.g. the condition given by Eq. (3.16). In Chapter 5, we also discussed the time stamps for decoding and presentation. In this chapter, we re-investigate conditions for preventing decoder buffer under-/over-flows for the constant delay channel from a slightly different view point by using the encoder timing, decoding time stamps and dynamics of encoded-picture size. We study some principles on video rate-buffer management of video encoders.

TV broadcast applications require that pictures input into encoder and output from decoder have the same frame (or picture) rate, and also require the video encoder and decoder to have the same clock frequency, and to operate synchronously. For example, in MPEG-2, decoder to encoder synchronization is maintained through the utilization of a program clock reference (PCR) and decoding time stamp (DTS) (or presentation time stamp (PTS)) in the bitstream. In an MPEG-2 transport stream, the adaptation field supplies a program clock reference (PCR). The PES packet supplies DTS and PTS. Since compressed pictures usually have variable sizes, DTSs (and PTSs) are related to encoder and decoder buffer (FIFO) fullness at certain points. Fig. 6.1 shows the video encoder and decoder buffer model. In this figure, T is the picture duration of the original uncompressed video as described in Chapter 3 and L is a positive integer. Thus, after a picture is encoded, it waits $L \cdot T$ before being decoded in the decoder.

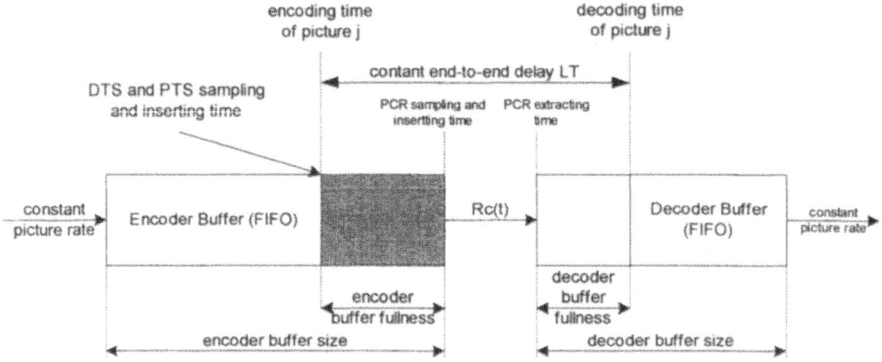

Figure 6.1 Video encoder and decoder buffer model.

The decoder buffer under- and/or over-flows are usually caused by channel jitter and/or video encoder buffer over- and/or under- flows. If the decoder buffer underflows, the buffer is being emptied faster than it is being filled. Coded bits resided in the decoder buffer are removed completely by the decoder at some point and some bits required by decoder are not yet received from the (assuming jitter-free) transmission channel. Consequently, too many bits are being generated in the encoder then at some point, i.e. the encoder buffer overflows. To prevent this, the following procedures are often used in the encoder at certain points:

- Increase the quantization level,
- Adjust bit-allocation,
- Discard high frequency DCT coefficients,
- Repeat pictures.

If the decoder buffer overflows, it is being filled faster than it is being emptied. Too many bits are being transmitted and too few bits are being removed by the decoder such that the buffer is full. Consequently, too few bits are being generated in the encoder at some point, i.e. encoder buffer underflows. To avoid this, the following procedures are often used in the encoder at certain point:

- Decrease the quantization level,
- Adjust bit-allocation,
- Stuff bits.

As shown in Chapter 3, the adjustments on quantization-level and bit-allocation are usually accomplished by using the rate-control algorithms along with an adaptive quantizer..

Rate-control and adaptive quantizer are important function blocks for achieving good compression performance in video encoder [6-2][6-3]. For this reason, every MPEG-2 encoding system in the market has its own intelligent rate-control and quantization algorithm. For example, every encoder has an optimized, and often complicated, bit-allocation algorithm to assign the number of bits for each type of pictures (I-, P-, and B-pictures). Such a bit-allocation algorithm usually takes into account the prior knowledge of video characters (e.g. scene changes, fade, etc.) and coding types (e.g. picture types) for a group of pictures (GOP). Adaptive quantization is applied in the encoder along with rate-control to ensure the required video quality and to satisfy the buffer regulation.

6.2 Conditions for Preventing Decoder Buffer Underflow and Overflow

The primary task of video rate-buffer management for an encoder is to control its output bit-stream to comply with the buffer requirements, e.g. the Video Buffering Verifier (VBV) specified in MPEG-2 video (ISO/IEC 13818-2)[6-1] and MPEG-4 video (ISO/IEC 14496-2) [6-2]. To accomplish such a task, rate-control algorithms are introduced in Chapter 3.

One of the most important goals for rate-control algorithms is to prevent video buffer under- and/or over-flows. For Constant Bit-Rate (CBR) applications, by a use of the rate-control, bit-count-per-second must precisely converge to the target bit-rate with good video quality. For Variable Bit-Rate (VBR) applications, the rate-control achieves the goal of maximizing the perceived quality of decoded video sequence with the maintained output-bitrate within permitted bounds.

In the following discussion, the encoder buffer is characterized by the following new set of parameters that are slightly different than those given in Chapter 3:

- B_E^j denotes the encoder buffer bit-level right before encoding of the j-th picture.
- b_D^j denotes the decoder buffer bit-level right before encoding of the j-th picture.
- P_E^j denotes the bit-count of the j-th coded picture.
- $B_D^{\max}$ denotes the decoder buffer size, e.g. MPEG-2 VBV buffer size coded in the sequence header and sequence extension if present.
- $B_E^{\max}$ denotes the size of the video encoder buffer.

Assume the encoding time of j-th picture is $t_{e,j}$ and decoding time of j-th picture is $t_{d,j}$, i.e. DTS for the j-th picture. Then, in order to avoid decoder buffer underflow, it requires that all the coded data up to and including picture j must be completely transmitted to the decoder buffer before time $t_{d,j}$, i.e.

$$B_E^j + P_E^j \leq \int_{t_{e,j}}^{t_{d,j}} R_C(t)dt, \qquad (6.1)$$

where $R_C(t)$ is the bit-rate function of the channel and the integral $\int_{t_{e,j}}^{t_{d,j}} R_C(t)dt$ represents the total of bits transmitted for the video service from time $t_{e,j}$ to $t_{d,j}$.

In order to avoid decoder buffer overflow, it requires that the decoder buffer fullness at time $t_{d,j}$ (before picture j is decoded) be less than $B_D^{\max}$. From time $t_{e,j}$ to $t_{d,j}$, the number of bits arriving at the decoder buffer will be

$\int_{t_{e,j}}^{t_{d,j}} R_C(t)dt$ and the number of bits being removed from the decoder buffer will be all the coded video data in both encoder and decoder buffers at time $t_{e,j}$. Thus the decoder buffer fullness at time $t_{d,j}$ satisfies:

$$b_D^j + \int_{t_{e,j}}^{t_{d,j}} R_C(t)dt - (b_D^j + B_E^j) \le B_D^{\max}. \tag{6.2}$$

This inequality can be simplified to

$$\int_{t_{e,j}}^{t_{d,j}} R_C(t)dt - B_E^j \le B_D^{\max}. \tag{6.3}$$

By applying inequality (6.3) to the (j+1)-th picture, one has

$$\int_{t_{e,j+1}}^{t_{d,j+1}} R_C(t)dt - B_E^{j+1} \le B_D^{\max}. \tag{6.4}$$

Where $t_{e,j+1}$ and $t_{d,j+1}$ denote the encoding and decoding time for picture j+1, respectively.

The encoder buffer fullness also satisfies the following recursive equation (which is similar to Eq.(3.7)) :

$$B_E^{j+1} = B_E^j + P_E^j - \int_{t_{e,j}}^{t_{e,j+1}} R_C(t)dt. \tag{6.5}$$

Thus, inequalities (6.4) and (6.5) yield

$$\int_{t_{e,j}}^{t_{d,j+1}} R_C(t)dt \le B_E^j + P_E^j + B_D^{\max}. \tag{6.6}$$

Inequalities (6.1) and (6.6) are necessary and sufficient conditions for preventing buffer under- and/or over- flows if they are held for all pictures.

By combining the two inequalities (6.1) and (6.6) one obtains upper and lower bounds on the size of picture j:

$$-B_D^{\max} + \int_{t_{e,j}}^{t_{d,j+1}} R_C(t)dt \le B_E^j + P_E^j \le \int_{t_{e,j}}^{t_{d,j}} R_C(t)dt \tag{6.7}$$

The above upper and lower bounds imply

$$-B_D^{\max} + \int_{t_{e,j}}^{t_{d,j+1}} R_C(t)dt \le \int_{t_{e,j}}^{t_{d,j}} R_C(t)dt . \tag{6.8}$$

This inequality (6.8) imposes a constraint on the transmission rate $R_C(t)$.

$$\int_{t_{d,j}}^{t_{d,j+1}} R_C(t)dt \le B_D^{\max} . \tag{6.9}$$

Also, from inequality (6.6), one has

$$\int_{t_{e,j}}^{t_{d,j+1}} R_C(t)dt - B_D^{\max} \le B_E^{\max} . \tag{6.10}$$

This inequality provides a lower bound on the encoder buffer size $B_E^{\max}$. Note that such a lower bound is determined by end-to-end (buffer) delay, transmission rate, and the decoder buffer size $B_D^{\max}$. This inequality is also consistent with the inequality (3.25) derived in Chapter 3.

Example:
In a MPEG-2 video transmission system, for an end-to-end (buffer) delay ($t_{d,j}-t_{e,j}=LT$, see Fig. 6.1) of 0.6 seconds, the time lag from $t_{e,j}$ to $t_{d,j+1}$ can be at most 0.6 seconds plus three field time (0.05 sec) in the case of 480i. Therefore, from inequalities (6.1) and (6.10), one has

$$B_E^{\max} \le 0.6 * R_C^{\max},$$

$$0.65 * R_C^{\max} - B_D^{\max} \le B_E^{\max},$$

where $B_D^{\max}$ = vbv_buffer_size. Thus, for $R_C^{\max}$ = 15000000 bits/second, the encoder buffer size is at most 1.125 Mbytes.

The video-buffer management protocol is an algorithm for checking a bitstream to verify that the amount of video-buffer memory required in the decoder is bounded by $B_D^{\max}$, e.g. vbv_buffer_size in MPEG-2 video. The rate-control algorithm will be guided by the video-buffer management protocol to ensure the bitstream satisfying the buffer regulation with good video quality.

One of the key steps in the video-buffer management and rate-control process is to determine the bit-budget for each picture. The condition given by inequality (6.1) on preventing the decoder buffer under-flow provides an upper bound on the bit-budget for each picture. The reason is that, at the decoding time, the current picture should be small enough so that it is contained entirely inside the decoder buffer. The condition given by inequality (6.6) on avoiding the decoder buffer over-flow provides a lower bound on the bit-budget for each picture. These conditions can also be directly applied to both MPEG-2 Test Model and MPEG-4 Verification Model rate-control algorithms shown in Chapter 3.

6.3 MPEG-2 Video Buffering Verifier

In MPEG-2 video, a coded video bitstream has to meet constraints imposed through a Video Buffering Verifier (VBV) defined in Annex C of reference [6-1]. The VBV is a hypothetical decoder, which is conceptually connected to the output of an encoder. It has an input buffer known as the VBV buffer (it is also called the rate buffer, sometimes). Coded data is placed in the buffer and is removed instantaneously at certain examination time from the buffer as defined in C.3, C.5, C.6, and C.7 of reference [6-1]. The time intervals between successive examination points of the VBV buffer are specified in C.9, C.10, C.11, and C.12. The VBV occupancy is shown in Fig. 6.2. It is required that a bitstream does not cause the VBV buffer to overflow. When there is no "skipped" picture, i.e. low_delay equals zero in MPEG-2 spec, the bitstream should not cause the VBV buffer to underflow.

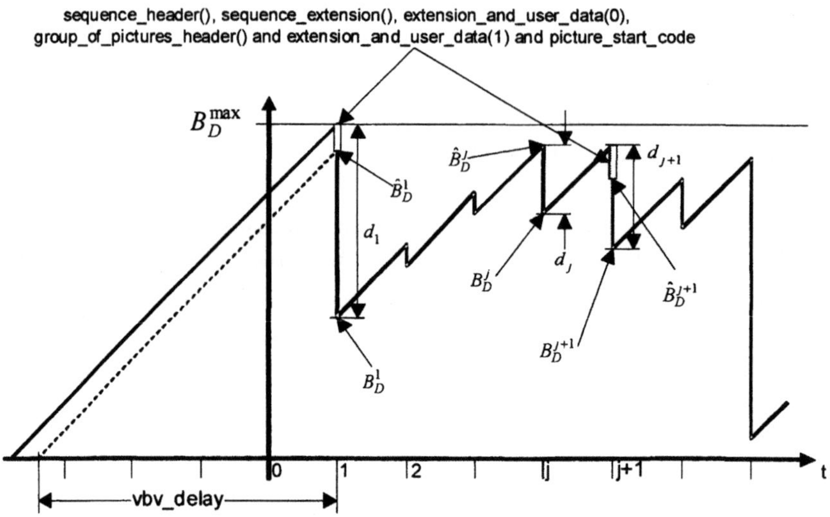

Figure 6.2 MPEG-2 VBV buffer occupancy (modified from [6-1]).

Thus, the condition for preventing the VBV buffer to overflow is

$$\hat{B}_D^j \le B_D^{max}.$$

And, the condition for preventing the VBV buffer to underflow is

$$0 \le B_D^j.$$

where B_D^{max} is the VBV buffer size. $\hat{B}_D^j$ is VBV occupancy, measured in bits, immediately before removing picture n from the buffer but after removing any header(s), user data and stuffing that immediately precedes the data elements of picture j. B_D^j is VBV occupancy, measured in bits, immediately after removing picture j from the buffer. Note that $\hat{B}_D^j - B_D^j = P_E^j = d_j$ is the size of the coded picture j and if the header bits can be ignore, then $\hat{B}_D^j = \hat{B}_D^{j-1} - d_{j-1} + R_C \cdot (t_j - t_{j-1})$.

In the constant bit-rate (CBR) case, $\hat{B}_D^j$ may be calculated by vbv_delay from the equation as follows:

$$\text{vbv\_delay}_\text{j} = \frac{90000\hat{B}_D^j}{R_C}, \tag{6.11}$$

i.e.,

$$\hat{B}_D^j = \frac{\text{vbv\_delay}_\text{j} \cdot R_C}{90000}, \tag{6.12}$$

where R_C denotes the actual bitrate (i.e. to full accuracy rather than the quantised value given by bit_rate in the sequence header). An approach to calculate the piecewise constant rate $R_C(j \cdot T)$ from a coded stream is specified in C.3.1 of reference [6-1].

Note that the encoder is capable of knowing the delay experienced by the relevant picture start code in the encoder buffer and the total end-to-end delay. Thus, the value encoded in vbv_delay (the decoder buffer delay of the picture start code) is calculated as the total end-to-end delay subtract the delay of the corresponding picture start code in the encoder buffer measured in periods of a 90 kHz clock derived from the 27 MHz system clock. Therefore, the encoder is able to generate a bitstream that does not violate the VBV constraints.

Initially, the VBV buffer is empty. The data input continues at the piecewise constant rate $R_C(j \cdot T)$. After filling the VBV buffer with all the data that precedes the first picture start code of the sequence and the picture start code itself, the VBV buffer is filled from the bitstream for the time specified by the vbv_delay field in the picture header. At this time decoding begins. By following this process (without looking the system's DTS), the decoder buffer will not over- and/or under-flow for VBV compliant streams.

Note that the ambiguity can happen at the first picture and the end of a sequence since input bit-rate cannot be determined from the bitstream. The ambiguity may become a problem when the video bitstream is remultiplexed and delivered at a rate different from the intended piecewise constant rate $R_C(j \cdot T)$.

For the CBR channel, if the initial vbv_delay$_0$ can be obtained, the decoding time can be determined from the vbv_delay$_0$ and picture (frame) rate T. For example, the decoding time for non-film mode video can be determined as follows:

$$t_{d,j} = \frac{90000 \cdot j}{T} + vbv\_delay_0 . \qquad (6.13)$$

In the variable bit-rate (VBR) case, i.e. vbv_delay is coded with the value hexadecimal FFFF, data enters the VBV buffer as specified as follows:

- Initially, the VBV buffer is empty.

- If the VBV buffer is not full, data enters the buffer at $R_C^{\max}$ where $R_C^{\max}$ is the maximum bit-rate specified in the bit_rate field of the sequence header.
- If the VBV buffer becomes full after filling at $R_C^{\max}$ for some time, no more data enters the buffer until some data is removed from the buffer.

This is, so called, the leak method since the video encoder for VBR transmission (including some transport buffers, see Chapter 8 for details) can be simply modeled as a leaky-bucket buffer, as described in section 3.3.2 of Chapter 3. In this case, $\hat{B}_D^j = \min(B_D^{\max}, \hat{B}_D^{j-1} - d_{j-1} + R_C^{\max} \cdot (t_j - t_{j-1}))$ if one ignores the header bits.

When there are skipped pictures, i.e. low_delay = 1, decoding a picture at the normally expected time might cause the VBV buffer to underflow. If this is the case, the picture is not decoded and the VBV buffer is re-examined at a sequence of later times specified in C.7 and C.8 of reference [6-1] until it is all present in the VBV buffer.

The VBV constraints ensure encoder buffer never over- and/or under-flow. A decoder that is built on a basis of VBV can always decompress the VBV compliant video streams without over- and/or under-flow the decoder buffer.

6.4 MPEG-4 Video Buffering Verifier

As discussed in the previous section, a video rate buffer model is required in order to bound the memory requirements for the bitstream buffer needed by a video decoder. With a rate buffer model, the video encoder can be constrained to make bit-streams that are decodable with a predetermined buffer memory size.

The MPEG-4 (ISO/IEC 14496-2) [6-2][6-10] video buffering verifier (VBV) is an algorithm for checking a bitstream with its delivery rate function, $R_C(t)$, to verify that the amount of rate buffer memory required in a decoder is less than the stated buffer size. If a visual bitstream is composed of multiple Video Objects (VO) and each VO is with one or more Video Object Layers (VOL), the rate buffer model is applied independently to each VOL (using buffer size and rate functions particular to that VOL). The concepts of VO, VOL and Video Object Plane (VOP) of MPEG-4 video are reviewed in Chapter 2.

In MPEG-4, the coded video bitstream is constrained to comply with the requirements of the VBV defined as follows:

- When the vbv_buffer_size and vbv_occupancy parameters are specified by systems-level configuration information, the bitstream shall be constrained according to the specified values. When the vbv_buffer_size and vbv_occupancy parameters are not specified (except for the short video header case for H.263 as described below), this indicates that the bitstream should be constrained according to the default values of vbv_buffer_size and vbv_occupancy. The default value of vbv_buffer_size is the maximum value of vbv_buffer_size allowed within the profile and level. The default value of vbv_occupancy is 170 × vbv_buffer_size, where vbv_occupancy is in 64-bit units and vbv_buffer_size is in 16384-bit units. This corresponds to an initial occupancy of approximately two-thirds of the full buffer size.
- The VBV buffer size is specified by the vbv_buffer_size field in the VOL header in units of 16384 bits. A vbv_buffer_size of 0 is forbidden. Define $B_D^{\max} = 16384 \times \text{vbv\_buffer\_size}$ to be the VBV buffer size in bits.
- The instantaneous video object layer channel bit rate seen by the encoder is denoted by $R_C(t)$ in bits per second. If the bit_rate field in the VOL header is present, it defines a peak rate (in units of 400 bits per second; a value of 0 is forbidden) such that $R_C(t) \leq R_C^{\max} = 400 \times \text{bit\_rate}$.
- The VBV buffer is initially empty. The vbv_occupancy field specifies the initial occupancy of the VBV buffer in 64-bit units before decoding the initial VOP. The first bit in the VBV buffer is the first bit of the elementary stream, except for basic sprite sequences.
- Define d_j to be size in bits of the j-th VOP plus any immediately preceding Group Of VOP (GOV) header, where j is the VOP index which increments by 1 in decoding order. A VOP includes any trailing stuffing code words before the next start code and the size d_j of a coded VOP is always a multiple of 8 bits due to start code alignment.
- Let $t_{d,j}$ be the decoding time associated with VOP j in decoding order. All bits d_j of VOP j are removed from the VBV buffer instantaneously at $t_{d,j}$. This instantaneous removal property distinguishes the VBV buffer model from a real rate buffer.

- The method of determining the value of $t_{d,j}$ is specified below. Assume $t_{p,j}$ is the composition time (or presentation time in a no-compositor decoder) of VOP *j*. For a VOP, $t_{p,j}$ is defined by vop_time_increment (in units of 1/vop_time_increment_resolution seconds) plus the cumulative number of whole seconds specified by module_time_base In the case of interlaced video, a VOP consists of lines from two fields and $t_{p,j}$ is the composition time of the first field. For example, the relationship between the composition time and the decoding time for a VOP is given by:

$$t_{d,j} = \begin{cases} t_{p,j}, & \text{if the j-th VOP is a B-VOP.} \\ t_{p,j} - m_j, & \text{otherwise.} \end{cases}$$

  In the normal decoding, the composition time of I and P VOP's is delayed until all immediately temporally-previous B-VOPs have been composed. This delay period is $m_j = t_{p,j} - t_{p,k}$, where *k* is the index of the nearest temporally-previous non-B VOP relative to VOP *j*.

  In order to initialize the model decoder when m_j is needed for the first VOP, it is necessary to define an initial decoding time $t_{d,0}$ for the first VOP (since the timing structure is locked to the B-VOP times and the first decoded VOP would not be a B-VOP). This defined decoding timing shall be that $t_{d,0} = 2 \cdot t_{d,1} - t_{d,2}$ (i.e., assuming that $t_{d,1} - t_{d,0} = t_{d,2} - t_{d,1}$), since the initial $t_{p,k}$ is not defined in the case.

The example given in Table 6.1 demonstrates how m_j is determined for a sequence with variable numbers of consecutive B-VOPs:

$$\text{Decoding order}: \quad I_0 P_1 P_2 P_3 B_4 P_5 B_6 P_7 B_8 B_9 P_{10} B_{11} B_{12}$$

$$\text{Presentation order}: \quad I_0 P_1 P_2 B_4 P_3 B_6 P_5 B_8 B_9 P_7 B_{11} B_{12} P_{10}$$

In this example, assume that *vop_time_increment=1* and *modulo_time_base=0* . The sub-index *j* is in decoding order.

- Define B_D^j as the buffer occupancy in bits immediately following the removal of VOP *j* from the rate buffer. Using the above definitions, B_D^j can be iteratively defined as

$$B_D^0 = 64 \cdot vbv\_occupancy - d_0$$

$$B_D^{j+1} = B_D^j + \int_{t_{d,j}}^{t_{d,j+1}} R_C(t)\cdot dt - d_{j+1} \text{ for } j \geq 0. \tag{6.14}$$

The rate buffer model requires that the VBV buffer never overflow or underflow, that is

$0 \leq B_D^j$ and $B_D^j + d_j \leq B_D^{\max}$ for all j.

Also, a coded VOP size must always be less than the VBV buffer size, i.e., $d_j \leq B_D^{\max}$ for all j. The MPEG-4 VBV buffer occupancy is shown in Fig. 6.3.

j	$t_{p,j}$	$t_{d,j}$	m_j
0	0	0-1=-1	1
1	1	1-1=0	1
2	2	2-1=1	1
3	4	4-2=2	2
4	3	3	2
5	6	6-2=4	2
6	5	5	2
7	9	9-3=6	3
8	7	7	3
9	8	8	3
10	12	12-3=9	3
11	10	10	3
12	11	11	3

Table 6.1 An example that demonstrates m_j is determined.

- If the short video header is in use (i.e., for H.263 baseline video [6-7][6-8]), then the parameter vbv_buffer_size is not present and the following conditions are required for VBV operation. The buffer is initially empty at the start of encoder operation (i.e., t=0 being at the time of the generation of the first video plane with short header), and its fullness is subsequently checked after each time interval of 1001/30000 seconds (i.e., at t=1001/30000, 2002/30000, etc.). If a complete video plane with short header is in the buffer at the examining time, it is removed. The buffer fullness after the removal of a VOP, B_D^j, shall be greater than or equal to zero and less than (4 · $R_C^{\max}$ · 1001) / 30000 bits, where $R_C^{\max}$ is the maximum bit rate in bits per second allowed within the profile and level. The number of bits used for coding any single VOP, d_j, shall not exceed $k \cdot 16384$ bits, where $k = 4$ for QCIF and Sub-QCIF, $k = 16$ for CIF, $k = 32$ for 4CIF, and $k = 64$ for 16CIF, unless a larger value of k is specified in the profile and level definition. Furthermore, the total buffer fullness at any time shall not exceed a value of $B_D^{\max} = k \cdot 16384 + (4 \cdot R_C^{\max} \cdot 1001) / 30000$.

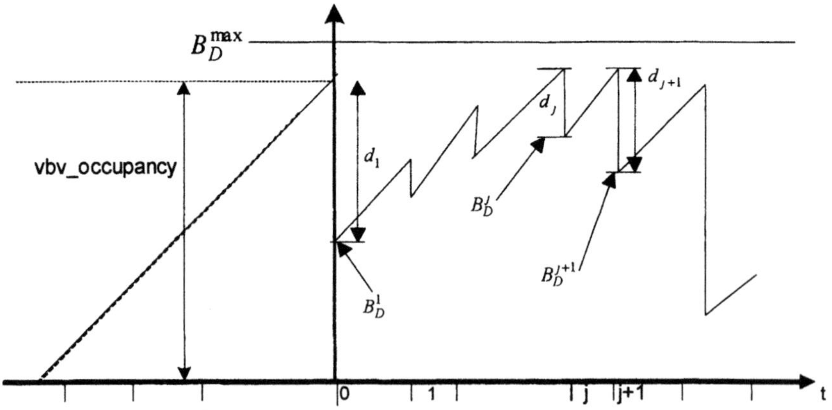

Figure 6.3 MPEG-4 VBV buffer occupancy.

It is a requirement on the encoder to produce a bitstream that does not overflow or underflow the VBV buffer. This means the encoder must be designed to provide correct VBV operation for the range of values of $R_C(t)$ over which the system will operate for delivery of the bitstream. A channel has constant delay if the encoder bit-rate at time t when particular bit enters the channel, the bit will be received at $t + LT$ and L is constant. In the case of constant delay channels, the encoder can use its locally estimated $R_C(t)$ to simulate the VBV occupancy and control the number of bits per VOP, $d_j = P_E^j$, in order to prevent overflow or underflow.

MPEG-4 VBV model assumes a constant delay channel. This allows the encoder to produce an elementary bitstream that does not overflow or underflow the buffer using $R_C(t)$.

6.5 Comparison between MPEG-2 VBV and MPEG-4 VBV

Both MPEG-2 and MPEG-4 VBV models [6-1] [6-10] specify that the rate buffer may not overflow or underflow and that coded pictures (VOPs) are removed from the buffer instantaneously. In both models a coded picture/VOP is defined to include all higher-level syntax immediately preceding the picture/VOP.

MPEG-2 video has a constant frame period (although the bitstream can contain both frame and field pictures and frame pictures can use explicit 3:2 pull-down via the repeat_first_field flag). In MPEG-4 terms, this frame rate would be the output of the compositor (the MPEG-2 terminology is the output of the display process that is not defined normatively by MPEG-2). This output frame rate together with the MPEG-2 picture_structure and repeat_first_field flag precisely defines the time intervals between consecutive decoded picture (either frames or fields) passed between the decoding process and the display process.

In general, the MPEG-2 bitstream contains B pictures (assume that low_delay = 0). This means the coding order and display order of pictures is different (since both reference pictures used by a B picture must precede the B picture in coding order). The MPEG-2 video VBV specifies that a B picture is decoded and presented (instantaneously) at the same time and the anchor pictures are re-ordered to make this possible. This is the same reordering model specified in MPEG-4 video.

A MPEG-4 model decoder using its VBV buffer model can emulate a MPEG-2 model decoder using the MPEG-2 VBV buffer model if the VOP time stamps given by vop_time_increment and the cumulative modulo_time_base agree with the sequence of MPEG-2 picture presentation times. Assume here that both coded picture/VOPs use the common subset of both standards (frame structured pictures and no 3:2 pulldown on the decoder, i.e., repeat_first_field = 0). For example, if the MPEG-4 sequence is coded at the NTSC picture rate 29.97Hz, vop_time_increment_resolution will be 30000 and the change in vop_time_increment between consecutive VOPs in presentation order will be 1001 because pictures are not allowed to skipped in MPEG-2 video when low_delay = 0.

MPEG-4 VBV does not specify the leaky bucket buffer model for VBR channel. However, the VBR model specified in MPEG-2 VBV can be applied to MPEG-4 video.

Bibliography

[6-1] ITU-T Recommendation H.262 | ISO/IEC 13818-2: 1995. Information technology – Generic coding of moving pictures and associated audio information: Video.
[6-2] ISO/IEC 14496-2:1998, Information Technology - Generic coding of audio-visual objects - Part 2: Visual.
[6-3] Test model editing committee, Test Model 5, MPEG93/457, ISO/IEC JTC1/SC29/WG11, April 1993.
[6-4] Xuemin Chen and Robert O. Eifrig, "Video rate buffer for use with push data flow", US Patent Number 6289129, Assignee: Motorola Inc. and General Instrument Corporation, Sept. 11, 2001.
[6-5] Atul Puri and T. H. Chen, Multimedia Standards and Systems, Chapman & Hall, New York, 1999.
[6-6] T.Sikora, "The MPEG-4 Video Standard Verification Model," IEEE Transactions on circuits and systems for video technology, Vol.7, No.1, Feb.1997.
[6-7] ITU-T Experts Group on Very Low Bitrate Visual Telephony, "ITU-T Recommendation H.263 Version 2: Video Coding for Low Bitrate Communication," Jan. 1998.
[6-8] ITU-T Experts Group on Very Low Bitrate Visual Telephony, "ITU-T Recommendation H.263: Video Coding for Low Bitrate Communication," Dec. 1995.
[6-9] B. G. Haskell, A. Puri, and A. N. Netravali, Digital Video: An Introduction to MPEG-2, New York: Chapman & Hall, 1997.
[6-10] Xuemin Chen and B. Eifrig, "Video rate buffer", ISO/IEC JTC1/SC29/WG11, M3596, July 1998.
[6-11] Xuemin Chen and Ajay Luthra, "A brief report on core experiment Q2—improved rate control", ISO/IEC JTC1/SC29/WG11, M1422 Maceio, Brizal, Nov. 1996.

[6-12] Xuemin Chen, B. Eifrig and Ajay Luthra, "Rate control for multiple higher resolution VOs: a report on CE Q2", ISO/IEC JTC1/SC29/WG11, M1657, Seville, Spain, Feb. 1997.

7 Transcoder Buffer Dynamics and Regenerating Timestamps

7.1 Video Transcoder

Digital video compression algorithms specified in the MPEG and H.263 standards [7-1] [7-2] [7-3] [7-9] [7-10] have already enabled many video services such as, video on demand (VoD), digital terrestrial television broadcasting (DTTB), cable television (CATV) distribution, and Internet video streaming. Due to the variety of different networks comprising the present communication infrastructure, a connection from the video source to the end user may be established through links of different characteristics and bandwidth.

In the scenario where only one user is connected to the source, or independent transmission paths exist for different users, the bandwidth required by the compressed video should be adjusted by the source in order to match the available bandwidth of the most stringent link used in the connection. For uncompressed video, this can be achieved in video encoding systems by adjusting coding parameters, such as quantization steps, whereas for pre-compressed video, such a task is performed by applying, so called, video transcoders [7-4], [7-5], [7-6], [7-11].

In the scenario where many users are simultaneously connected to the source and receiving the same coded video, as happen in VoD, CATV services and Internet video, the existence of links with different capacities poses a serious problem. In order to deliver the same compressed video to all users, the source has to comply with the sub-network that has the lowest available capacity. This unfairly penalizes those users that have wider bandwidth in their own access links. By using transcoders in communication links, this problem can be resolved. For a video network with transcoders in its subnets, one can ensure that users receiving lower quality video are those having lower bandwidth in their transmission paths. An example of this scenario is in CATV services where a satellite link is used to transmit compressed video from the source to a ground station, which in turn distributes the received video to several destinations through networks of different capacity.

In the scenario where the compressed video programs need to be re-assembled and re-transmitted, the bit rates of the coded video are often reduced in order to fit in the available bandwidth of the channel. For example, cable head-ends can re-assemble programs from different video sources. Some programs from broadcast television and others from video servers. In order to ensure that the re-assembled programs can match the available bandwidth, video transcoders are often used.

When a transcoder is introduced between an encoder and the corresponding decoder, the following issues should be considered for the system [7-4][7-5][7-6][7-11]:

- Buffer and delay.
- Video decoding and re-encoding.
- Timing recovery and synchronization.

In Chapter 2, many video compression technologies are discussed. As an extension, two basic video transcoder architectures are overviewed here. Transcoding is an operation of converting a pre-compressed bit stream into another bit stream at different rate. For example, a straightforward architecture of transcoder for MPEG bit stream can simply be a cascaded MPEG decoder/encoder [7-11], as shown in Fig. 7.1. In the cascaded-based transcoder, the pre-compressed MPEG bit stream is first decompressed by the cascaded decoder and the resulting reconstructed video sequence is then re-encoded by the cascaded encoder, which generates a new bit stream. The desired rate of the new bit stream can often be achieved by adjusting quantization level, Q_2, in the cascaded encoder. The main concern with the cascaded-based transcoder is its implementation cost: one full MPEG decoder and one full MPEG encoder.

Recent studies showed that a transcoder consisting of a cascaded decoder/encoder can be significantly simplified if the picture types in pre-compressed bit stream can remain unchanged during transcoding [7-4][7-6][7-11], that is, a decoded I-picture is again coded as an I-picture, a decoded P-picture is again coded as a P-picture and a decoded B-picture is again coded as a B-picture. In fact, by maintaining the picture types, one can possibly reduce the complexity of motion estimation (ME) (the most expensive operation) by using small search around decoded motion vectors (MVs) (as shown in the dished-line in Fig. 7.1).

One can also remove ME in the cascaded-based transcoder (Fig. 7.1) because of the fact that there is a strong similarity between the original and the reconstructed video sequences. Hence, a MV field that is good for an original coded picture should be reasonably good for the corresponding re-encoded picture.

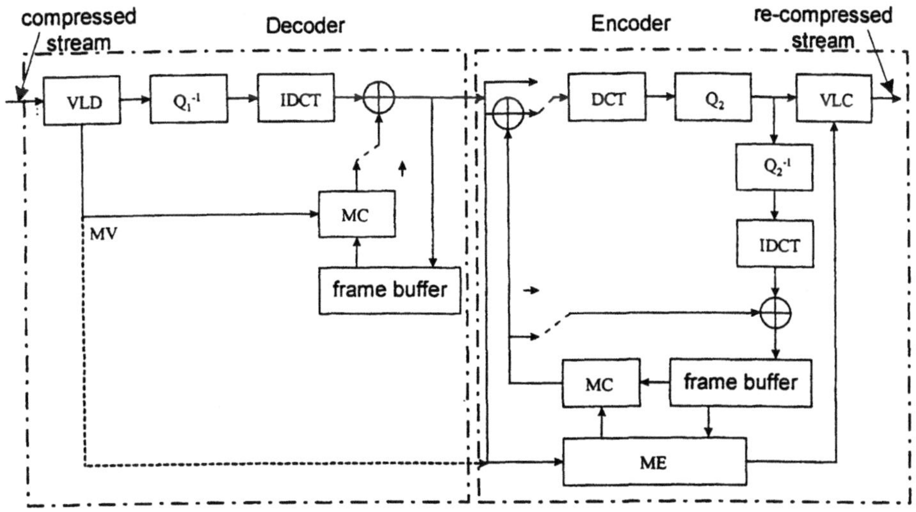

Figure 7.1. Cascaded MPEG decoder/encoder.

Fig. 7.2 shows a cascaded-based transcoder without ME where the MV fields required for MC in the cascaded encoder are now obtained from the cascaded decoder. However, it should also be pointed out that although the MV fields obtained from the cascaded decoder can be reasonably good for motion compensation (MC) in the cascaded encoder (Fig. 7.2), they are not the best because they were estimated based upon the original coded sequence. For example, the half-pixel positions of re-used MVs could be inaccurate.

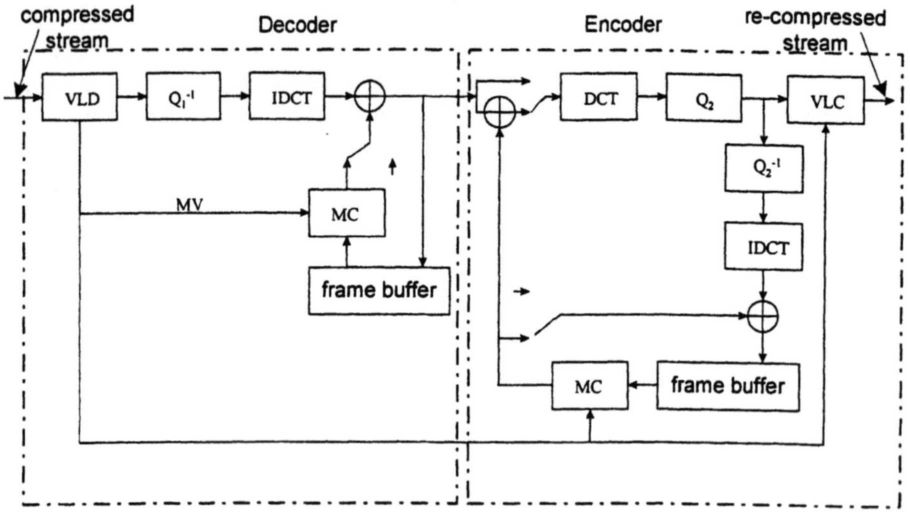

Figure 7.2. Cascaded MPEG decoder/encoder without ME where MV required for MC in the cascaded encoder is obtained from the cascaded decoder.

Many other transcoder architectures [7-5][7-6] can be derived or extended from the two basic architectures given in Figs. 7.1 and 7.2. For example, a transcoder with picture resolution change is developed in [7-6].

In the remaining of the chapter, the discussion will be focused on analyzing buffer, timing recovery and synchronization for video transcoder. The buffering implications of the video transcoder within the transmission path are analyzed. For transcoders with either fixed or variable compression ratio, it is shown that the encoder buffer size can be maintained as if no transcoder existed while the decoder has to modify its own buffer size according to both the bit rate conversion ratio and transcoder buffer size. The buffer conditions of both the encoder and transcoder are derived for preventing the decoder buffer from underflowing or overflowing. It is also shown that the total buffering delay of a transcoded bit stream can be made less than or equal to its "encoded-only" counterpart.

The methods for regenerating timestamps for transcoder are also introduced in this chapter.

7.2 Buffer Analysis of Video Transcoders

Smoothing buffers play a important role in transmission of coded video. Therefore, if a transcoder is introduced between an encoder and the corresponding decoder, some modifications are expected to be required in the existing buffering arrangements of a conventional encoder-decoder only system, which is primarily defined for being used without transcoders.

It is known that encoders need an output buffer because the compression ratio achieved by the encoding algorithm is not constant throughout the video signal. If the instantaneous compression ratio of a transcoder could be made to follow that of the encoder, then no smoothing buffer would be necessary at the transcoder [7-8]. For a CBR system this requires a fixed-transcoding compression ratio exactly equal to the ratio between the output and input CBR bit rates of the transcoder. In general, this is impossible to obtain in practice and a small buffer is necessary to smooth out the difference.

In the following analysis, the general case of buffer dynamics is first presented and then, in order to clarify the effect of adding a transcoder in the transmission path, the cases of fixed compression ratio transcoding without buffering, and variable compression ratio transcoding with buffering, are analyzed. The concept of video data unit is usually defined as the amount of coded data that represents an elementary portion of the input video signal such as block, macroblock, slice or picture (a frame or a field). In the following analysis, the video (data) unit is assumed to be a picture and the processing delay of a video unit is assumed to be constant in either the encoder, transcoder or decoder and is much smaller than the buffering delays involved. Thus, this delay can be neglected in the analysis model. For the same reasons, the transmission channel delay is also neglected in the analysis model. A video unit is characterized by the instant of its occurrence in the input video signal, as well as by the bit rate of the corresponding video data unit. Since the processing time is ignored, video units are instantly encoded into video data units, and these then instantly decoded into video units. Although video data units are periodically generated by the encoder, their periodicity is not maintained during transmission (after leaving the encoder buffer) because, due to the variable compression ratio of the coding algorithm, each one comprises a variable number of bits. However, for real-time display, the decoder must recover the original periodicity of the video units through a synchronized clock.

Buffer dynamics of the encoder-decoder only system: [7-8] Before analyzing buffering issues in transcoders, let us look again at the relationship between encoder and decoder buffers in a codec without transcoding such as the general case of transmission depicted in Figure. 6.1 in Chapter 6, where the total delay $L \cdot T$ from the encoder input (e.g. camera) to the decoder output (e.g. display) is the same for all video units (T is the picture duration of the original uncompressed video as described in Chapter 3). Therefore, since processing and transmission delays are constant, a video data unit entering into the encoder buffer at time t will leave the decoder buffer at $t + L \cdot T$ where $L \cdot T$ is constant. Since the acquisition and display rates of corresponding video units are equal, the output bit rate of the decoder buffer at time $t + L \cdot T$ is exactly the same as that of the input of the encoder buffer at time t. Thus, in Fig. 6.1, assume that $E_0(t)$ represents the bit rate of a video data unit encoded at time t and the coded video data is transmitted at a rate $R_C(t)$. If underflow or overflow never occurs in either the encoder or decoder buffers, then the encoder buffer fullness $B_E(t)$ is given by Eq. (7.1), while that of the decoder buffer $B_D(t)$ is given by Eq. (7.2)

$$B_E(t) = \int_0^t E_0(\tau) \cdot d\tau - \int_0^t R_C(\tau) \cdot d\tau \tag{7.1}$$

$$B_D(t) = \begin{cases} \int_0^t R_C(\tau) \cdot d\tau, & t < L \cdot T \\ \int_0^t R_C(\tau) \cdot d\tau - \int_0^{t - L \cdot T} E_0(\tau) \cdot d\tau, & t \geq L \cdot T \end{cases} \tag{7.2}$$

Note that in general, it is possible for encoder buffer under-flow to occur if transmission starts at the same time as the encoder puts the first bit into the buffer, as implied by Eq. (7.1). In practice, this is prevented by starting the transmission only after a certain initial delay such that the total system delay is given by the sum of the encoder and decoder initial delays, L_e and L_d, respectively. For simplicity, one can assume that encoder buffer underflow does not occur and then these two delays are included in the total initial decoding delay, i.e. $L \cdot T = L_e \cdot T + L_d \cdot T$. From Eq. (7.2), it can be seen that during the initial period $0 \le t < L \cdot T$ the decoder buffer is filled at the channel rate up to the maximum $\int_0^{L\cdot T} R_C(\tau) \cdot d\tau$, hence decoding of the first picture only starts at $t = L \cdot T$. Combining Eqs. (7.1) and (7.2) yields that the sum of the encoder and decoder buffer occupancies at times t and $t + L \cdot T$ respectively, is bounded and equal to the buffer size required for the system, i.e.,

$$\begin{aligned} & B_E(t) + B_D(t + L \cdot T) \\ & = \int_0^t E_0(\tau) \cdot d\tau - \int_0^t R_C(\tau) \cdot d\tau + \int_0^{t+L\cdot T} R_C(\tau) \cdot d\tau - \int_0^t E_0(\tau) \cdot d\tau \\ & = \int_t^{t+L\cdot T} R_C(\tau) \cdot d\tau \le L \cdot T \cdot R_C^{\max}, \quad t \ge L \cdot T \end{aligned} \tag{7.3}$$

For a VBR channel, the sum of buffer occupancies of both encoder and decoder is the total amount of bits that have been transmitted from *(t, t+LT)*. Then, in this case, the above equation shows that, for a constant delay channel, the buffer size required for the system is $L \cdot T \cdot R_C^{\max}$, where $R_C^{\max}$ is the maximum channel rate. For a CBR channel, one has $R_C^{\max} = R_C^{\min} = \text{constant}$ where $R_C^{\min}$ is the minimum channel rate. Then, the above equation also shows that the total number of bits stored in both the encoder and decoder buffers at any times t and $t + L \cdot T$, respectively, is always the same, i.e. $B_E(t) + B_D(t + L \cdot T) = R \cdot L \cdot T$. Thus, if these bits "travel" at a CBR, the delay between encoder and decoder is maintained constant for all video units while the sum of buffer occupancies of both encoder and decoder is a constant for all video units. . The principal requirement for the encoder is that it must control its buffer occupancy such that decoder buffer overflow or underflow never occurs. Decoder buffer overflow implies loss of data whenever its occupancy reaches beyond the required buffer size $B_D^{\max}$. On the other hand, underflow occurs when the decoder buffer occupancy is zero at display time of a video unit that is not fully decoded yet (display time is externally imposed by the display clock).

Eq. (7.3) relates encoder and decoder buffer occupancies at time t and $t + L \cdot T$, respectively. This buffer fullness equation provides the conditions for preventing encoder and decoder buffers being over- or under-flow. Decoder buffer underflow is prevented if $B_D(t + L \cdot T) \ge 0$ is ensured at all times. Thus, using Eq. (7.3), at time t the encoder buffer fullness should be

$$B_E(t) \leq \int_t^{t+L\cdot T} R_C(\tau)\cdot d\tau. \tag{7.4}$$

On the other hand, decoder buffer overflow does not occur if

$$B_D(t+L\cdot T) \leq \int_t^{t+LT} R_C(\tau)\cdot d\tau \tag{7.5}$$

holds all the time, which requires that the encoder buffer occupancy at time meets the following condition: $B_E(t) \geq 0$.

Therefore, it can be seen that decoder buffer underflow and overflow can be prevented by simply controlling the encoder buffer occupancy such that $0 \leq B_E(t) \leq \int_t^{t+L\cdot T} R_C(\tau)\cdot d\tau$ at any time t. By preventing the encoder buffer from overflowing, its decoder counterpart never underflows while preventing encoder buffer underflow ensures that the decoder buffer never overflows. More buffering requirements can be found in Chapters 3 and 6.

Inequalities (7.4) and (7.5) also imply that the maximum needed encoder and decoder buffer sizes satisfy: $B_E^{\max} = B_D^{\max} = L\cdot T\cdot R_C^{\max}$. This means that the specified buffer size for either encoder or decoder needs no more than $L\cdot T\cdot R_C^{\max}$.

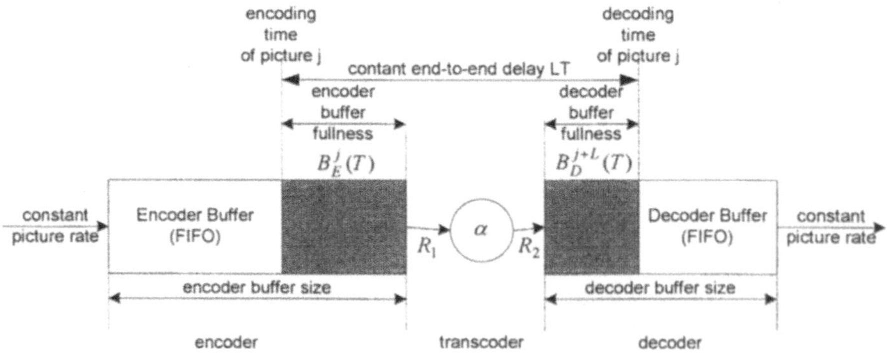

Figure 7.3 A transcoder model with fixed-transcoding compression ratio.

The MPEG-2 standard defines both VBR and CBR transmission. The sequence header of an MPEG-2 bit stream includes the decoder buffer size and the maximum bit rate that can be used. Also, in each picture header is included the time e.g. vbv_delay, that the decoder should wait after receiving the picture header until start decoding the picture. For CBR transmission vbv_delay is such that the encoder and decoder buffer dynamics are as explained above and the total delay is kept constant. It is calculated by the encoder as the difference (in number of periods of the system clock) between the total delay and the delay that each picture header undergoes in the encoder buffer.

Transcoder with a fixed compression ratio: Next, consider the CBR transmission case. Let us now assume that a hypothetical transcoder, capable of achieving a fixed compression ratio transcoding of $\alpha = \frac{R_2}{R_1}$ such that $0 < \alpha \leq 1$ is inserted in the transmission path as shown in Fig. 7.3. R_1 and R_2 are the input and output CBR's of the transcoder, respectively. The bit-rate that enters into the encoder buffer is reduced through the factor α such that the output of the decoder buffer is a delayed and scaled version of $E_0(t)$, given by $\alpha \cdot E_0(t - L \cdot T)$. Because of the lower channel rate at the decoder side, if the total delay is to be kept the same as if no transcoder was used, then the decoder buffer fullness level is usually lower than that of the decoder without transcoder being used in the system. The encoder assumes a normal CBR transmission without transcoders in the network thus, if any of the system parameters encoded in the original bit stream such as bit rate, buffer size, and vbv_delay in the headers of MPEG-2 video bit streams need to be updated, the transcoder has to perform the task in a transparent manner with respect to both the encoder and decoder. The initial delay is set up by the decoder as waiting time before decoding the first picture. Hence, if neither buffer underflows nor overflows, the encoder and decoder buffer occupancies are given by

$$B_E(t) = \int_0^t E_0(\tau) \cdot d\tau - R_1 \cdot t \tag{7.6}$$

$$B_D(t) = \begin{cases} R_2 \cdot t, & t < L \cdot T \\ R_2 \cdot t - \int_0^{t-L\cdot T} \alpha \cdot E_0(\tau) \cdot d\tau, & t \geq L \cdot T \end{cases} \tag{7.7}$$

Using Eqs. (7.6) and (7.7), it can be shown that the delay $L \cdot T$ between encoder and decoder is maintained constant at any time t. In this case, a video data unit entering into the encoder buffer at time t will leave the decoder buffer at time $t+L\cdot T$, hence the sum of the waiting times in both the encoder and decoder buffers is given by

$$\begin{aligned}
&\frac{1}{R_1}\cdot B_E(t)+\frac{1}{R_2}\cdot B_D(t+L\cdot T)\\
&=\frac{1}{R_1}\int_0^t E_0(\tau)\cdot d\tau-\frac{1}{R_1}\int_0^t R_1\cdot d\tau+\frac{1}{R_2}\int_0^{t+L\cdot T} R_2\cdot d\tau-\frac{\alpha}{R_2}\int_0^t E_0(\tau)\cdot d\tau\\
&=\frac{1}{R_1}\int_0^t E_0(\tau)\cdot d\tau-t+t+L\cdot T-\frac{1}{R_2}\int_0^t \alpha\cdot E_0(\tau)\cdot d\tau\\
&=L\cdot T
\end{aligned} \tag{7.8}$$

since $\alpha=\frac{R_2}{R_1}$, the above equation shows that the total delay from encoder to decoder is still the same constant regardless of the transcoder being inserted along the transmission path. However, because the encoder and decoder buffers work at different CBR's, the sum of the buffer occupancies $B_E(t)+B_D(t+L\cdot T)$ is no longer constant as was in the previous case of transmission without the transcoder. Since the input bit rate of the decoder buffer is lower than the output bit rate of the encoder buffer, for the given end-to-end delay $L\cdot T$, the maximum needed encoder and decoder buffer sizes, $B_E^{\max}$ and $B_D^{\max}$, respectively, can be derived as follows. Then, from Eq. (7.8), one has

$$\begin{aligned}
&\frac{B_E(t)}{R_1}\le L\cdot T\le\frac{B_E(t)}{R_1}+\frac{B_D^{\max}}{R_2},\\
&\frac{B_D(t)}{R_2}\le L\cdot T\le\frac{B_E^{\max}}{R_1}+\frac{B_D(t)}{R_2}
\end{aligned}. \tag{7.9}$$

Thus,

$$\begin{aligned}
&L\cdot T\cdot R_1\ge B_E(t)\ge L\cdot T\cdot R_1-\frac{R_1}{R_2}\cdot B_D^{\max},\\
&L\cdot T\cdot R_2\ge B_D(t)\ge L\cdot T\cdot R_2-\frac{R_2}{R_1}\cdot B_E^{\max}.
\end{aligned} \tag{7.10}$$

By definition, one has $0\le B_E(t)\le B_E^{\max}, 0\le B_D(t)\le B_D^{\max}$. Thus, it is known from inequality (7.9) that the maximum needed buffer sizes satisfy

$$L\cdot T=\frac{B_E^{\max}}{R_1}=\frac{B_D^{\max}}{R_2}. \tag{7.11}$$

i.e.

$$B_D^{\max} = \frac{R_2}{R_1} \cdot B_E^{\max} = \alpha \cdot B_E^{\max} . \quad (7.12)$$

Eq. (7.12) shows that by using a smaller decoder buffer with size $B_D^{\max} = \alpha \cdot B_E^{\max}$, the same total delay can be maintained as if no transcoder existed.

Let us now analyze the implications of the small decoder buffer size on the encoder buffer constraints needed to prevent decoder buffer underflow and overflow. Assuming that the encoder is not given any information about the transcoder then, recalling the case of CBR transmission without transcoders, the encoder prevents decoder buffer overflow and underflow by always keeping its own buffer occupancy within the limits $0 \le B(t) \le B_E^{\max}$. With a similar approach to Eq. (7.8), the system delay is

$$\frac{1}{R_1} \cdot B_E(t) + \frac{1}{R_2} \cdot B_D(t + L \cdot T) = L \cdot T \quad (7.13)$$

where it can be seen that decoder buffer underflow never occurs if at display time $t + L \cdot T$ all the bits of the corresponding video data unit are received, i.e., $B_D(t + L \cdot T) \ge 0$ after removing all its bits from the buffer. Hence, using Eqs. (7.11) and (7.13)

$$B_D(t + L \cdot T) = B_D^{\max} - \frac{R_2}{R_1} \cdot B_E(t) \ge 0 , \quad (7.14)$$

and the condition for preventing decoder buffer underflow is given by

$$B_E(t) \le \frac{R_1}{R_2} \cdot B_D^{\max} = B_E^{\max} . \quad (7.15)$$

On the other hand, decoder buffer does not overflow if its fullness is less than the buffer size immediately before removing all the bits of any video data unit, i.e., hence, using again Eqs. (7.11) and (7.13)

$$\frac{R_2}{R_1} \cdot B_E(t) \ge 0 \quad (7.16)$$

since $\frac{R_2}{R_1} \ge 0$, then decoder buffer overflow is prevented, providing that

$$B_E(t) \ge 0 . \quad (7.17)$$

Inequalities (7.15) and (7.17) show that no extra modification is needed at the encoder for preventing decoder buffer underflow or overflow. By controlling the occupancy of its own buffer of size $B_E^{\max}$ such that overflow and underflow never occurs, the encoder is automatically preventing the smaller decoder buffer from underflowing and overflowing. This means that, in this case, the presence of the transcoder can be simply ignored by the encoder without adding any extra buffer restrictions on the decoder. In this case, an MPEG-2 transcoder would have to modify the buffer size $B_E^{\max}$ specified in its incoming bit stream to a new value $\frac{R_2}{R_1} \cdot B_E^{\max}$ while the delay parameter in picture headers should not be changed because the buffering delay at the decoder is exactly the same as in the case where no transcoder is used.

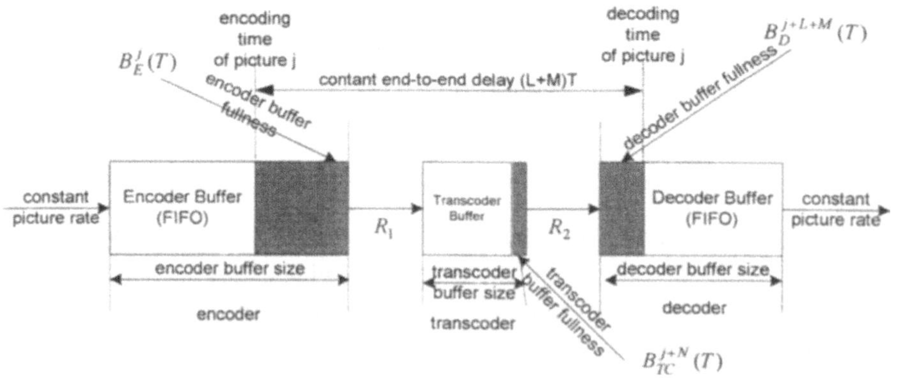

Figure 7.4 A transcoder model with variable-transcoding compression ratio.

However, a transcoder with a fixed compression ratio as was assumed in this case is almost impossible to obtain in practice, mainly because of the nature of the video-coding algorithms and the compressed bit streams they produce. Such a transcoder would have to output exactly $\alpha \cdot N$ bits for each incoming bits N. Since each video data unit consists of a variable number of bits and the quantized DCT blocks cannot be finely encoded such that a given number of bits is exactly obtained, a perfectly fixed compression ratio transcoder cannot be implemented in practice. Moreover, a transcoder with variable compression ratio may be even desirable if the objective is, for instance, to enforce a given variable transcoding function. The above analysis of a fixed compression ratio transcoder provides relevant insight into the more practical case to be described next.

Transcoder with a Variable Compression Ratio: As was pointed out before, a transcoder with variable compression ratio must incorporate a smoothing buffer in order to accommodate the rate change of the coded stream. The conceptual model of a CBR transmission system including such a transcoder with a local buffer of size $B_{TC}^{\max}$ is illustrated in Fig. 7.4. The encoder buffer size $B_E^{\max}$ is maintained as in the previous cases while that of the decoder should be given by $B_D^{\max} = \alpha \cdot B_E^{\max} + B_{TC}^{\max}$, as shall be explained later. Here, transcoding is modeled as a scaling function $\beta(t)$ which multiplied by $r(t) = E_0(t)$ produces the transcoded VBR $r'(t)$, i.e.,

$$r'(t) = \beta(t) \cdot r(t) \,. \tag{7.18}$$

The effect of multiplying $\beta(t)$ by $r(t)$ can be seen as equivalent to reducing the number of bits used in the video data unit encoded at time *t*. The output of the decoder buffer consists of a delayed version of $r'(t)$. In the system of Fig. 7.4, transcoding is performed on the CBR R_1 which consists of the video data units of $r(t)$ after the encoder buffering delay $L_e(t)$, defined as the delay that a video data unit encoded at time waits in the encoder buffer before being transmitted.

Let us now verify that under normal conditions, where neither of the three buffers underflows or overflows, the total delay between encoder and decoder is still a constant $M \cdot T + L \cdot T$ where $M \cdot T$ is the extra delay introduced by the transcoder. A video data unit entering the encoder buffer at time *t* will arrive at the transcoder at $t + L_e(t)$ and will be decoded in the final decoder at $t + M \cdot T + L \cdot T$, respectively. Since the processing delay in the transcoder is neglected, $t + L_e(t)$ is also the time at which the video data unit is transcoded and put in the transcoder buffer. Therefore, in order to calculate the total delay of the system, the encoder, transcoder and decoder buffers should be analyzed at instants *t*, $t + L_e(t)$, and $t + M \cdot T + L \cdot T$, respectively. The following Eqs. (7.19)-(7.21) provide the encoder buffer occupancy at time *t*, the transcoder buffer occupancy at time $t + L_e(t)$, and the decoder buffer occupancy at time $t + M \cdot T + L \cdot T$, respectively

$$B_E(t) = \int_0^t E_0(\tau) \cdot d\tau - R_1 \cdot t \tag{7.19}$$

$$B_{TC}(t + L_e(t)) = \int_0^t E_0(\tau) \cdot d\tau - (t + L_e(t)) \cdot R_2 \tag{7.20}$$

$$B_D(t + (M + L) \cdot T) = (t + (M + L) \cdot T) \cdot R_2 - \int_0^t r'(\tau) \cdot d\tau \tag{7.21}$$

A video data unit entering the encoder buffer at time t has to wait for $L_e(t)=\frac{B_E(t)}{R_1}$ seconds before leaving this buffer, plus $\frac{B_{TC}(t+L_e(t))}{R_2}$ in the transcoder buffer before being transmitted to the decoder buffer from which it is finally removed at $t+M\cdot T+L\cdot T$. Using the above equations for the buffer occupancies, the total buffering delay L_{TOTAL} from encoder to decoder is given by

$$\begin{aligned} L_{TOTAL} &= \frac{1}{R_1}\cdot B_E(t)+\frac{1}{R_2}\cdot B_{TC}(t+L_e(t))+\frac{1}{R_2}\cdot B_D(t+(M+L)\cdot T) \\ &= \frac{1}{R_1}\int_0^t E_0(\tau)\cdot d\tau - t \\ &+\frac{1}{R_2}\int_0^t r'(\tau)\cdot d\tau-(t-L_e(t)) \\ &+\frac{1}{R_2}\left((t+(M+L)\cdot T)\cdot R_2-\int_0^t r'(\tau)\cdot d\tau\right). \end{aligned} \tag{7.22}$$

where $L_e(t)=\frac{1}{R_1}\int_0^t E_0(\tau)\cdot d\tau - t$.

By simplifying the above expression, one has

$$\begin{aligned} L_{TOTAL} &= L_e(t)+\frac{1}{R_2}\int_0^t r'(\tau)\cdot d\tau - t - L_e(t) \\ &+t+(M+L)\cdot T-\frac{1}{R_2}\int_0^t r'(\tau)\cdot d\tau \\ &=(M+L)\cdot T \end{aligned} \tag{7.23}$$

It can be seen that the total delay is constant as given by the initial decoding delay. Note that, similar to the case of a transcoder with fixed compression ratio, the sum of the occupancies of the three buffers is not constant because of the different CBR's involved.

Since the encoder is assuming a decoder buffer of size $L\cdot T\cdot R_1$ (from Eq. (7.3)), its own buffer occupancy is kept within the limits $0\le B(t)\le B_E^{\max}$, as was shown earlier, is necessary for preventing decoder buffer overflow and underflow. However, since the actual required size of the decoder buffer is $B_D^{\max}$, the constraints that the transcoder buffer should meet in order to prevent decoder buffer underflow and overflow are derived from the system delay Eq. (7.24),

$$\frac{B_D^{\max}}{R_2} = \frac{1}{R_1} \cdot B_E(t) + \frac{1}{R_2} \cdot B_{TC}(t + L_e(t)) + \frac{1}{R_2} \cdot B_D(t + (M + L) \cdot T) = L_{TOTAL} \quad (7.24)$$

and substituting $B_D^{\max} = \alpha \cdot B_E^{\max} + B_{TC}^{\max}$ where $\alpha = \frac{R_2}{R_1}$, the decoder buffer occupancy is given by

$$B_D(t + (M + L) \cdot T) = \alpha \cdot B_E^{\max} + B_{TC}^{\max} - \alpha \cdot B_E(t) - B_{TC}(t + L_e(t)) \quad (7.25)$$

To ensure the decoder buffer not underflowing one has that

$$B_D(t + (M + L) \cdot T) \geq 0,$$

i.e.

$$\alpha \cdot B_E^{\max} + B_{TC}^{\max} - \alpha \cdot B_E(t) - B_{TC}(t + L_e(t)) \geq 0. \quad (7.26)$$

This is equivalent to constrain the transcoder buffer occupancy such that

$$B_{TC}(t + L_e(t)) \leq \alpha \cdot B_E^{\max} + B_{TC}^{\max} - \alpha \cdot B_E(t) \quad (7.27)$$

Since $B_E(t) \leq B_E^{\max}$, the decoder buffer never underflows if the transcoder buffer fullness is constrained such that

$$B_{TC}(t) \leq B_{TC}^{\max} \quad (7.28)$$

Similarly, $B_D(t + (M + L) \cdot T) \leq \alpha \cdot B_E^{\max} + B_{TC}^{\max}$ is the condition that the decoder buffer should meet for not overflowing. Thus, using Eq.(7.25), one obtains

$$\alpha \cdot B_E^{\max} + B_{TC}^{\max} - \alpha \cdot B_E(t) - B_{TC}(t + L_e(t)) \leq \alpha \cdot B_E^{\max} + B_{TC}^{\max} \quad (7.29)$$

which is equivalent to constrain the transcoder buffer occupancy, such that

$$-\alpha \cdot B_E(t) \leq B_{TC}(t + L_e(t)) \,. \quad (7.30)$$

Hence, in order to prevent the decoder buffer from overflowing, it is sufficient that the following condition holds all the time

$$B_{TC}(t) \geq 0 \,. \quad (7.31)$$

In summary, for a constant delay channel, if both the encoder and transcoder buffers never underflow or overflow, then decoder buffer will never overflow or underflow. The basic idea is that by increasing the total delay between encoder and decoder, a buffer corresponding to this extra delay can be used in the transcoder, which in turn is responsible for preventing it from overflowing and underflowing. Therefore, the encoder assumes a decoder buffer of size $L \cdot T \cdot R_1$ and the decoder is informed that the encoder is using a buffer size $B_D^{\max} = \alpha \cdot B_E^{\min} + B_{TC}^{\min}$. Between them, the transcoder performs the necessary adaptation, such that the process is transparent for both the encoder and decoder. For MPEG-2 video, the transcoder is responsible for updating the buffer size specified in the sequence header, as well as the delay parameter of each picture header. For MPEG-4 video, the transcoder is also responsible for updating the buffer size as well as the initial VBV occupancy specified in the Video Object Layer (VOL) header.

7.3 Regenerating Time Stamps in Transcoder

As studied in section 7.1, a transcoder involves decoding and re-encoding processes. Thus, the idea method for re-generating PCR, PTS and DTS is to use a phase-lock loop for the video transport stream. Fig. 7.5 shows a model of re-generating time stamps in transcoder.

In this model, assume that the encoding time of the transcoder can be ignored. PCR_E denote PCRs being inserted in the encoder while PCR_T, PTS_T, and DTS_T denote time stamps being re-inserted in the transcoder. STC_T is the new system time-base (clock) for the transcoder. In this case, the entire timestamp-insertion process is similar to that in the encoder.

For many applications, it is too expensive to have a PLL in the transcoder for each video program, especially in a multiple channel transcoder [7-5]. Instead, the transcoder can use one free running system clock (assuming that the clock is accurate, e.g., exactly 27MHz) and perform PCR correction. An example of the PCR correction is shown in Fig. 7.6.

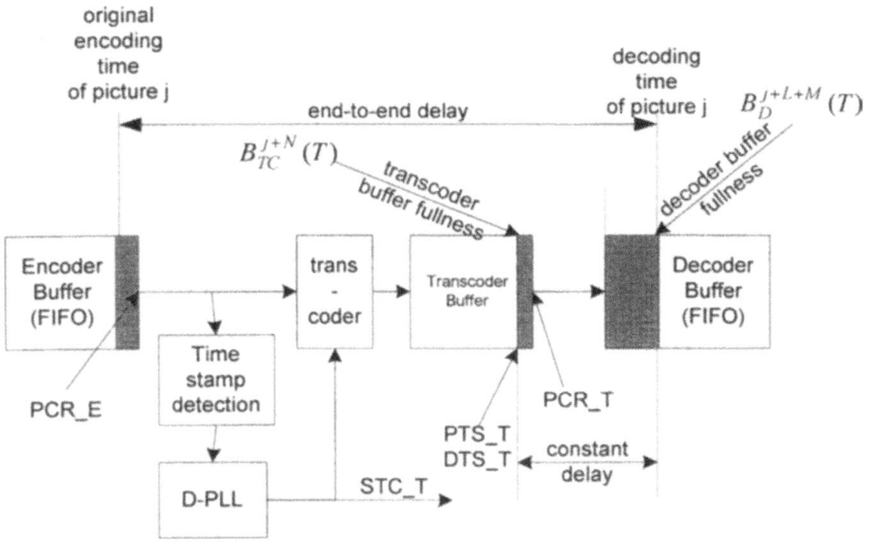

Figure 7.5 A model of regenerating time stamps in transcoder.

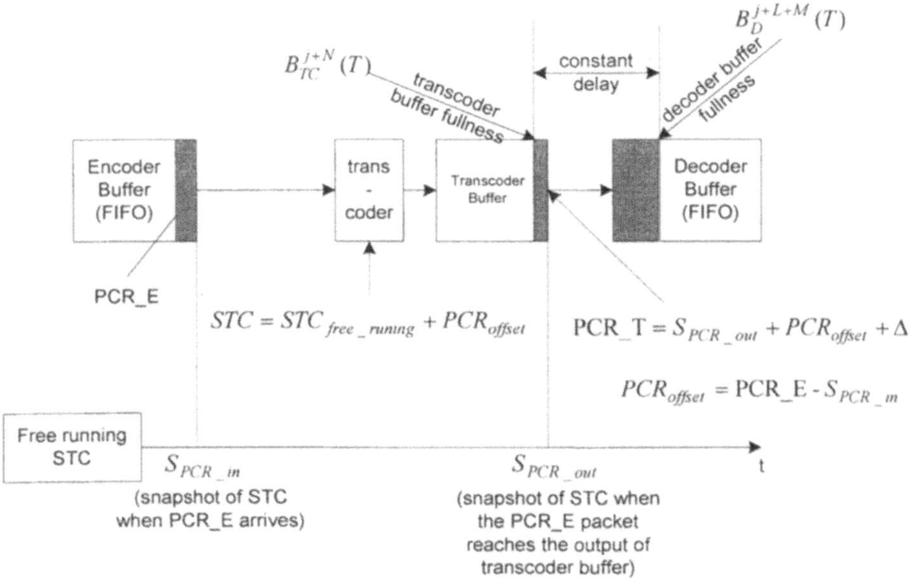

Figure 7.6 A model of the PCR correction

In Fig. 7.6, one free-running system time clock, STC_{free_runing} is used (for all channels). When a TS packet with a PCR (PCR_E) arrives, the snapshot value S_{PCR_in} of $STC_{free\_running}$ is taken and the difference between PCR_E and S_{PCR_in} is computed as $PCR_{offset} = PCR\_E - S_{PCR\_in}$. Then, the instantaneous system time clock for the channel is

$$STC = STC_{free\_running} + PCR_{offset}. \tag{7.32}$$

The snapshot of the time when the same PCR packet reaches the output of the transcoder buffer can also be taken as S_{PCR_out}. Then, the new PCR value for the transcoder output can be generated by

$$\text{PCR\_T} = S_{PCR\_out} + PCR_{offset} + \Delta, \tag{7.33}$$

where Δ is an estimated error due to small difference between the transcoder free-running clock counter and the video encoder STC counter.

Both PTS and DTS values for the transcoder can be generated in a similar manner. One can also keep the original PTS and DTS values, but only adjust PCR_T by subtracting the delay between the transcoder decoding time and the final decoder decoding time.

Bibliography

For books and articles devoted to video transcoding coding systems :

[7-1] Ralf Schafer and Thomas Sikora, "Digital video coding standards and their role in video communications", Proceeding of IEEE, Vol. 83, No. 6, pp.907-924, June 1995.

[7-2] ITU-T Recommendation H.262 (1995) | ISO/IEC 13818-2: 1996, Information technology – Generic coding of moving pictures and associated audio information: Video.

[7-3] ISO/IEC 14496-2:1998, Information Technology - Generic coding of audio-visual objects - Part 2: Visual.

[7-4] G. Keesman, R. Hellinghuizen, F. Hoeksema, and G. Heideman, "Transcoding of MPEG bitstreams," Signal Processing: Image Communication, vol.8, pp.481-500, Sept. 1996.

[7-5] Xuemin Chen and Fan Ling, "Implementation architectures of a multi-channel MPEG-2 video transcoder using multiple programmable processors", US Patent No. 6275536B1, Aug. 14, 2001.

[7-6] Xuemin Chen, Limin Wang, Ajay Luthra, Robert Eifrig, "Method of architecture for converting MPEG-2 4:2:2-profile bitstreams into main-profile bitstreams", US Patent No. 6259741B1, July 10, 2001.
[7-7] Xuemin Chen, Fan Lin, and Ajay Luthra, "Video rate-buffer management scheme for MPEG transcoder", WO0046997, 2000.
[7-8] P. A. A. Assuncao, and M. Ghanbari, "Buffer analysis and control in CBR video transcoding", IEEE Tans. On Circuit and Systems for Video Technology, vol.10, No. 1, Feb. 2000.
[7-9] ITU-T Experts Group on Very Low Bitrate Visual Telephony, "ITU-T Recommendation H.263 Version 2: Video Coding for Low Bitrate Communication," Jan. 1998.
[7-10] ITU-T Experts Group on Very Low Bitrate Visual Telephony, "ITU-T Recommendation H.263: Video Coding for Low Bitrate Communication," Dec. 1995.
[7-11] L. Wang, A. Luthra, and B. Eifrig, "Rate-control for MPEG transcoder", IEEE Trans. On Circuit and Systems for Video Technology, vol. 11, No. 2, Feb. 2001.

8 Transport Packet Scheduling and Multiplexing

8.1. MPEG-2 Video Transport

The MPEG-2 transport stream is overviewed in this section as an example of typical video transport mechanisms. Some terminologies are also defined in here for discussion in Chapters 8 and 9.

Transport Stream coding structure: MPEG-2 transport stream (TS) [8-1] allows one or more programs to be combined into a single stream. Video and audio elementary streams (ES) are multiplexed together with information that allows synchronized presentation of these ES within a program. Video and audio ES consist of access units. Usually, the video access unit is a coded picture while the audio access unit is a coded audio frame.

Each video and audio ES is carried in PES packets. A PES packet consists of a PES packet header followed by payload. PES packets are inserted into TS packets. The PES packet header begins with a 32-bit start-code that also identifies the stream or stream type to which the packet data belongs. The PES packet header carries decoding and presentation time stamps (DTS and PTS). The PES packet payload has variable length.

A TS packet, as already discussed in Chapter 1, begins with a 4-byte prefix, which contains a 13- bit Packet ID (PID). The PID identifies, via the Program

Specific Information (PSI) tables, the contents of the data contained in the TS packet. Two most important PSI tables are

- Program Association Table (PAT).
- Program Map Table (PMT).

These tables contain the necessary and sufficient information to de-multiplex and present programs. The PMT specifies, among other information, which PIDs, and therefore which elementary streams are associated to form each program. This table also indicates the PID of the TS packets that carry the PCR for each program. TS packets may be null packets that are intended for padding of TS. These null packets may be inserted or deleted by re-multiplexing processes.

Transport Stream System Target Decoder (T-STD): The basic requirements for specifying a video transport standard are

- To generate packets of coded audio, video, and user-defined private data and,
- To incorporate timing mechanisms to facilitate synchronous decoding and presentation of these data at the client side.

In MPEG-2 standard, these requirements led to the definition of the Transport System Target Decoder (T-STD) [8-1]: The Transport System Target Decoder is an abstract model of an MPEG decoding terminal that describes the idealized decoder architecture and defines the behavior of its architectural elements. The T-STD provides a precise definition of time and recovery of timing information from information encoded within the streams themselves, as well as mechanisms for synchronizing streams with each other. It also allows the management of decoder's buffers.

The T-STD model consists of a small front-end buffer (with size of 512 bytes) called the transport buffer (TB) that receives the TS packets for the video or audio stream from a specific program identifier (PID) and that outputs the received TS packets at a specified rate. The output stream of a TB is sent to the decoder main buffer that is drained at times specified by the decoding time stamps (DTSs).

There are three types of decoders in the T-STD: video, audio, and systems. A diagram of video T-STD model is shown in Figure 8.1.

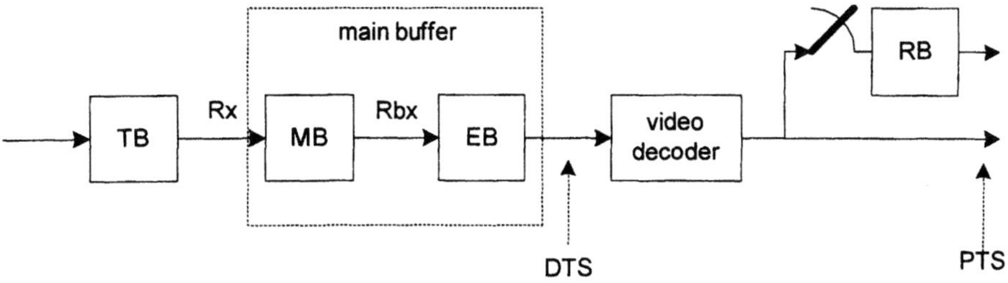

Figure 8.1 MPEG-2 video T-STD model.

In Fig. 8.1, TB denotes the transport buffer for a video ES. The main buffer consists of two buffers: the multiplexing buffer MB of the video ES and the video ES buffer EB. RB denotes the frame re-ordering buffer.

Timing information for the T-STD is carried by several data fields defined in [8-1]. These data fields carry two types of timestamps:

- Program clock references (PCRs) are samples of an accurate bitstream-source system clock (system clock frequency is 27Mhz). A MPEG decoder feeds PCRs to a phase-locked loop to recover an accurate time-base synchronized with the bitstream source.
- Decoding time stamps (DTSs) and presentation time stamps (PTSs) tell a decoder when to decode and when to present (display) compressed video pictures and audio frames.

Input to the T-STD is a TS. A TS may contain multiple programs with independent time bases. However, the T-STD decodes only one program at a time. In the T-STD model all timing indications refer to the time base of that program.

Data from the Transport Stream enter the T-STD at a piecewise constant rate. The time at which this byte enters the T-STD can be recovered from the input stream by decoding the input PCR fields, encoded in the Transport Stream packet adaptation field of the program to be decoded and by counting the bytes in the TS between successive PCRs for the program to be decoded. The PCR is encoded in two parts [8-1]: the first one, in units of the period of 1/300 times the system clock frequency (yielding 90 kHz), is called program_clock_reference_base, and the second one is called program_clock_reference_ext in units of the period of the system clock frequency.

In normal case, i.e. there is no time-base discontinuity, the transport rate is determined as the number of bytes in the Transport Stream between the bytes containing the last bit of two successive PCR fields of the same program divided by the difference between the time values encoded in these same two PCR fields.

TS packets containing data from the video ES, as indicated by its PID, are passed to the TS buffer for the stream. This includes duplicate TS packets and packets with no payload. All bytes that enter the buffer TB are removed at the rate Rx specified below.

$$R_X = \begin{cases} 0, & \text{if TB is empty,} \\ 1.2 \cdot \mathrm{R}_{\max}(\text{profile, level}), & \text{Otherwise.} \end{cases}$$

$\mathrm{R}_{\max}(\text{profile, level})$ is specified in table 8-13 of [8-1] according to the profile and level, e.g. $\mathrm{R}_{\max}(\text{main, main}) = 15\text{Mbits/s}$ for the main profile at the main level of MPEG-2 video. TB cannot overflow and must empty at least once every second. This imposes restrictions on input rate $R_{TS}(t)$ of the TS:

$$\int_0^t R_{TS}(\tau) \cdot d\tau - R_X \cdot t \le 512 \times 8 = 4096 \text{ bits for all } t,$$

and there exists t_1 and t_2 such that

$$t_2 - t_1 \le 1 \text{ second}$$

and $\int_{t_1}^{t_2} R_{TS}(\tau) \cdot d\tau - R_X \cdot (t_2 - t_1) = 0$.

The size of MB is defined as

$$\mathrm{MBS} = \mathrm{BS}_{\mathrm{mux}} + \mathrm{BS}_{\mathrm{oh}} + \begin{cases} \mathrm{VBV}_{\max}(\text{profile, level}) - \text{vbv\_buffer\_size}, & \text{for low and main level.} \\ 0, & \text{for high1440 and high level.} \end{cases}$$

where $\mathrm{VBV}_{\max}(\text{profile, level})$ is defined in table 8-12 of [8-1]. where $\mathrm{BS}_{\mathrm{oh}}$, PES packet overhead buffering is defined as:

$$\mathrm{BS}_{\mathrm{oh}} = \frac{1}{750} \cdot \mathrm{R}_{\max}(\text{profile, level}),$$

and $\mathrm{BS}_{\mathrm{mux}}$, additional multiplex buffering is defined as:

$$\mathrm{BS}_{\mathrm{mux}} = 0.004 \cdot \mathrm{R}_{\max}(\text{profile, level}).$$

The ES buffer size is defined for video as equal to the vbv_buffer_size as it is carried in the sequence header. EB cannot underflow except when the low delay flag in the video sequence extension is set to '1' (6.2.2.3 of [8-1]) or trick_mode status is true.

MPEG-2 systems standard [8-1] specifies one of two methods, the leak method or the VBV delay method, being used for transferring video data from MB to EB. When the leak method is in used, MB cannot overflow, and must become empty at least once every second. When the vbv_delay method is used, MB cannot overflow nor underflow, and EB cannot overflow.

Elementary stream buffered in EB is decoded instantaneously by video decoder and may be delayed in reorder buffers RB before being presented to the viewer at the output of the T-STD. Reorder buffers are used only in the case of a video elementary stream when some access units are not carried in presentation order. These access units will need to be reordered before presentation. In particular, if a picture is an I-picture or a P-picture carried before one or more B-pictures, then it must be delayed in the reorder buffer, RB, of the T-STD before being presented.

8.2 Synchronization in MPEG-2 by using STD

Synchronization in MPEG-2 is handled at the transport and PES layers, with the PCR, PTS and DTS fields serving as instruments. After the incoming transport stream is de-multiplexed into individual video and audio TS packets in the input queues of their corresponding STD, the video and audio PESs are extracted from TS packets, which are then forwarded to their respective decoders. The decoders parse the PES headers to extract the PTS and DTS fields. Note that PTS and DTS fields are not necessarily encoded for each video picture or audio presentation unit, but are only required to appear with intervals not, exceeding 0.7 second for periodic updating of the decoders' clocks. Whereas the DTS's specify the time at which all the bytes of a media presentation unit are removed from the buffers for decoding, the PTS's specify the actual time at which the presentation units are displayed to the user.

The STD model assumes instantaneous decoding of media presentation units. For audio units and B pictures of video, the decoding time is the same as the presentation time, and so only their PTS's are listed in their respective PES headers. On the other hand, I- and P-pictures of video have a reordering delay (since their transmission and decoding would have preceded earlier B pictures -- see Chapter 5) between their decoding and presentation, and hence, their PTS and DTS values differ by some integral number of picture (or field) periods (equal to the display time of the earlier B pictures). After the PTS and DTS are extracted from PES header for a media presentation unit, the data bytes are routed for decoding and display.

In some applications such as picture-in-picture (PIP) and a recorded video program being playback from the disk, the display of different media units must proceed in a mutually synchronized manner. The synchronization can be driven by one of the video steams serving as the master.

Synchronization Using a Master Stream: In this approach, all of the media streams being decoded and displayed must have exactly one independent master [8-4]. Each of the individual media display unit must slave the timing of their operation to the master stream. The master stream may be chosen depending on the application. Whichever media stream is the master, all the media streams but the master must slave the timing of their respective displays to the PTS's extracted from the master media steam.

To illustrate the approach, assume that the audio stream is chosen to be the master; the audio playback will drive the progression of playback of all the steams. The audio stream will be played back continuously with the clock being continually updated to equal the PTS value of the audio unit being presented for display. In particular, the STD clock is typically initialized to be equal to the value encoded in the first PCR field when that field enters the decoder's buffer. Thereafter the audio decoder controls the STD clock. As the audio decoder decodes audio presentation units and displays them, it finds PTS fields associated with those audio presentation units. At the beginning of display of each presentation unit, the associated PTS field contains the correct value of the decoder's clock in an idealized decoder following the STD model. The audio decoder uses this value to update the clock immediately.

The other decoders simply use the audio-controlled clock to determine the correct time to present their decoded data, at the times when their PTS fields are equal to the current value of the clock. Thus, video units are presented when the STD clock reaches their respective PTS values, but the clock is never derived from a video PTS value. Therefore, if the video decoder lags for any reason, it may be forced to skip presentation of some video pictures. On the other hand, if the video decoder leads, it may be forced to pause (repeat to display the previous picture). But the audio is never skipped or paused -- it will always proceed at its natural rate since the audio has been chosen to be the master.

For most of the MPEG-2 transport applications, synchronization is directly driven from PCR by a timing-recovery circuit. In general sense, this can be called synchronization in distributed playback.

Synchronization in Distributed Playback: [8-4] In this case, the PCR derived system clock serves as the time master (STD clock), with the audio and video decoders implemented as separate decoder subsystems, each receiving the complete multiplexed stream or the TS stream for the PID. Each decoder parses the received stream and extracts the system layer information and the coded data needed by that decoder. The decoder then determines the correct time to start decoding by comparing the DTS field extracted from the stream with the current value of its STD clock.

In the idealized STD, audio and video decoding is assumed to be instantaneous. Real decoders, however, may experience nonzero decoding delays; furthermore, the audio and video decoding delays may be different, causing them to go out of synchrony. Over a period of one second (i.e., 90000 cycles), errors of 50 parts per million can lead to PTS values differing from the nominal values by 4 or 5, which accumulates over time. In order to maintain proper synchronization, the timing-recovery circuit must track the real-time progression of playback at the decoders, for which purpose, the decoders transmit feedback messages to the timing-recovery circuit. A feedback message can be a light-weight packet that is transmitted concurrently with the display of a media unit, and contains the PTS of that media unit. When a feedback message from a decoder arrives at the timing-recovery circuit, the timing-recovery circuit extracts the PTS contained in the feedback. PTS's extracted from feedbacks of different decoders, when compared, reveal the asynchrony if any, for example, by muting the leading stream until the lagging stream catches up.

This synchronization approach can also be used in the case that the video and audio decoders are physically at different locations on the network. In this case neither video nor audio can be used as the master.

8.3 Transport Packet Scheduling

An MPEG-2 Transport Stream may be comprised of one or more services. Each of these services is formed by one or more components that have a common time-base. A service encoder will generate a Transport Stream that is made up of n different services. Each service has a single component PID stream that will carry the PCR for the service. Transport Streams generated by several service encoders may be combined by the Packet Multiplexer to create a single Transport Multiplex.

To create a "legal" Transport Stream, the following requirements are specified for service encoders:

- The Transport Stream must be generated at a rate that will not result in overflow or underflow of any buffer in the Transport Stream System Target Decoder (T-STD). The T-STD is a conceptual model described in the MPEG-2 Systems standard.
- The service encoder must multiplex transport packets created from several elementary streams into a single packet stream. Each of these elementary streams will generate transport packets at different rates. The service encoder should schedule packets in the multiplex so that the packet rate of each elementary stream is maintained within the Transport Stream with minimum multiplexing error.
- The Program Clock Reference (PCR) field must be sent periodically in one of the elementary streams in each service. The time interval between successive occurrences of the PCR in the Transport Stream must be less than or equal to 0.1 seconds.

Figure 8.2 shows the model used by the Packet Scheduler to create the Transport Stream in a service encoder.

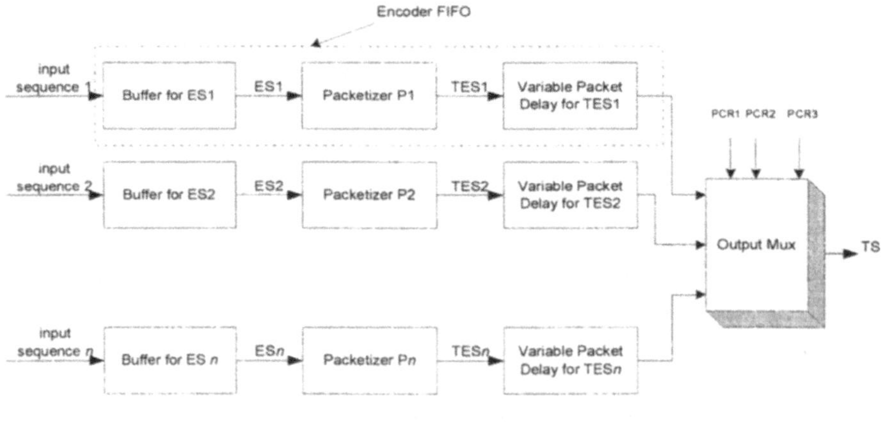

Figure 8.2 Packet Scheduler Model

The MPEG-2 transport encoder will deliver 188-byte transport packets at a constant rate, R_{TS}. The Transport Stream carries *n* programs, each made up of one or more of the *m* component, or elementary, streams. The component streams could be:

- Audio
- Video
- Isochronous Data
- etc..

The video, audio, and isochronous data streams will be formed into packetized elementary streams (PES). These PES streams are held in separate buffers prior to transport.

The model shown in Fig. 8.2 has a packetizer *Pj* assigned to each of the *m* elementary streams. *Pj* will read out component stream (*ESj*) data from buffer *j* and create a stream of 188-byte transport packets of constant rate, R_j. *ESj* may also be formed into a PES stream at this point if the component is video, audio, or isochronous data. The packet rate, R_{TS}, of the transport encoder output is the sum of the packet rates of all *m* PES transport streams, that is,

$$R_{TS} = \sum_{j=1}^{m} R_j \ .$$

For each time *t* that a transport packet must be sent on the Transport Stream, the Packet Scheduler will select a packet from those awaiting transport in the *m* packet delay blocks. The selected packet will be the one that has the least time in the packet delay blocks before the next packet originating from the same elementary stream emerges from the packetizer. In other words, the Packet Scheduler will evaluate

$$\min\{t_1(i_1+1)-t, t_2(i_2+1)-t, \ldots\ldots, t_m(i_m+1)-t\},$$

where

$t_1(i_1) \le t < t_1(i_1+1)$ and $t_1(i_1)$ is currently in the delay block 1,

$t_2(i_2) \le t < t_2(i_2+1)$ and $t_2(i_2)$ is currently in the delay block 2,

$t_m(i_m) \le t < t_m(i_m+1)$ and $t_m(i_m)$ is currently in the delay block m.

Using this method for packet selection ensures that any single packet will be sent in the Transport Stream before the next packet originating from the same component stream is ready for transport. As a result, the amount of time

$td_j(i)$ that packet $TES_j(i)$ is delayed in its packet delay block is less than the time interval $t_j(i+1)-t_j(i)$ between $TES_j(i)$ and $TES_j(i+1)$, i.e.

$$td_j(i) < t_j(i+1)-t_j(i) = \frac{1}{R_j}.$$

Each of the n services that are sent on the transport stream may have a different time base. A value derived from the appropriate time base will be sent periodically in the PCR field of a component stream with a specified PID that is assigned to a given program. For example, assume that there are two services that contain four elementary streams, ES_1, ES_2, ES_3, and ES_4. If ES_2 is assigned to Service 1, and ES_1, ES_3 and ES_4 are assigned to Service 2, transport packets from ES_2 will periodically include a Service 1 PCR and ES_1 transport packets will periodically include a Service 2 PCR. Fig. 8.2 also shows the point PCRs being inserted into the Transport Stream.

Figure 8.3 is an example of how packet scheduling would be performed to assemble a Transport Stream from three component streams of different rates. Each box represents a 188-byte transport packet. In the example, TES_1 and TES_3 are both assigned to Service 1, with TES_1 carrying the Service 1 PCR field. TES_1 is assigned to Service 2, and its packets carry the Service 2 PCR field.

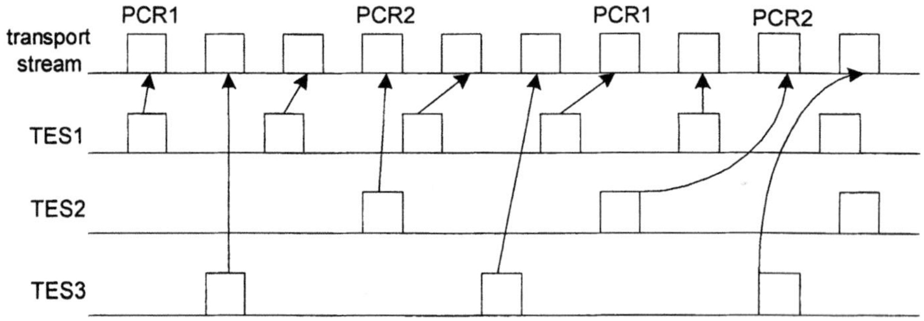

Figure 8.3. Transport Stream Packet Scheduling

8.4 Multiplexing of Compressed Video Streams

Technologies of multiplexing several variable rate encoded video stream into a single stream are discussed in this chapter. These technologies can be applied in satellite or cable video transmission, multimedia presentations with multiple video streams, and video on demand.

In digital video services, such as the satellite or cable digital television, video and audio encoders are often co-located while the associated decoders may or may not be co-located. In these applications, a fixed number of different video channels is encoded and transmitted together, and bit-rate for each channel can be controlled by a central multiplexing unit. When more than one stream are multiplexed, it is essential that data is not lost by encoder or decoder buffer overflow or underflow. One straightforward solution is to increase the buffer size in the system. However not only is this inefficient, it may not solve the problem, especially if the system has a variable transmission or retrieval rate.

The MPEG-1, MPEG-2 and MPEG-4 audio-video coding standards [8-1], [8-2],[8-3] support multiplexing mechanisms for combining bit-streams from up to 32 audio, 16 video, many video objects and any number of auxiliary streams. The channel rate used for transmission or retrieval from storage need not be constant, but may be variable. Therefore, transmission may be across a leased line or across a packet-switched public network, for example. Alternatively, retrieval could be from a DVD-ROM database that has a bursty data rate. However, implementation architectures of multiplexing are not provided in these standards.

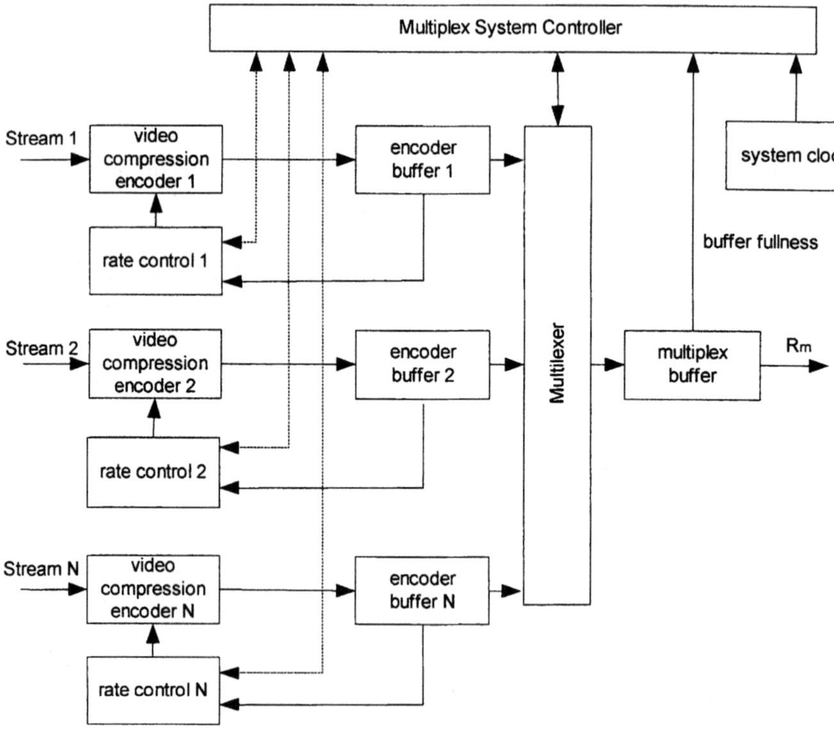

Figure 8.4 A multiplexing system

In this section, we describe an implementation model whereby multiple encoded bitstreams can be multiplexed into a single bitstream (of either constant or variable rate) such that encoder, decoder, and multiplex buffers do not overflow or underflow. To facilitate editing of the stored multiplexed streams, it is specifically required in this model that the parts of the individual streams that were generated during the same time interval be the number of individual sources is constant and known prior to multiplexing. At the de-multiplexer, it is also assumed that each decoder has its own buffer and that there is no buffering prior to multiplexing. This allows, for example, easy integration of any number of video, audio, and data decoders into very flexible configurations. Also, A rate control at the encoders is required to prevent overflow and underflow of encoder and decoder buffers.

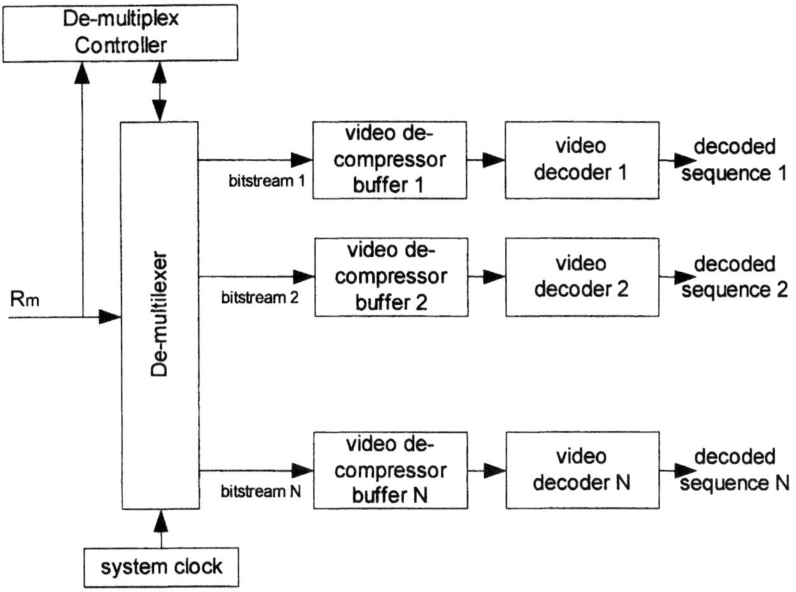

Figure 8.5 De-multiplexer system 1

A Model of Multiplexing Systems: The transport multiplexing system is shown in Fig. 8.4 for combining multiple streams into a single bit-stream of rate *Rm* bits/second. Initially assume that each encoder has a small buffer of its own, and multiplexed stream is fed to a much larger multiplex buffer prior to transmission by the channel. If the demultiplexer were a mirror image of the multiplexer, i.e. large demultiplex buffer prior to demultiplexing, then the system would be fairly straightforward as described in [8-1]-[8-4]. However, in many applications independent decoders (including buffers) are utilized as shown in Fig. 8.5. An even simpler arrangement is possible, as shown in Figure 8.6, if each decoder is able to identify and extract its own packets of data from the multiplexed bit-stream. In this case, additional decoders can be added, as designed, simply by connecting them to the incoming data.

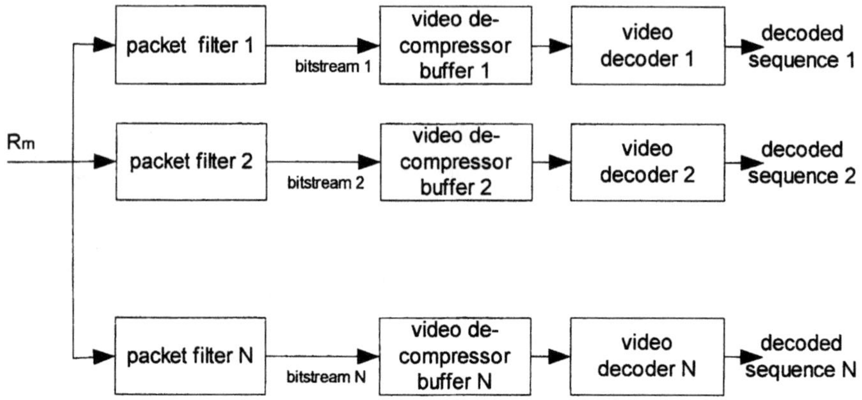

Figure 8.6 De-multiplexing system 2.

Next, the system model given in Figure 8.4 is described in more detail. Several media streams, labeled 1, 2,... enter from the left. Each stream consists of a sequence of access units. For video, an access unit comprises the bits necessary to represent a single coded picture (e.g. a frame in MPEG-1 and MPEG-2 or a video object plane in MPEG-4). For audio, an access unit could be a block of samples. Assume that each stream has assigned to it some nominal average bit rate, and that each encoder endeavors to operate near its assigned rate using perhaps the methods of [8-5]. Note that burstiness is allowed if there is sufficient channel capacity. However, buffer overflow may threaten if too many sources transmit above their assigned rates for too long.

Consider for now stream 1. Access units from stream 1 enter the first encoder where they are encoded into one or more packets of data and fed to its encoder buffer. The start and end times of each access unit as well as the number of bits generated during coding are monitored by encoder rate control 1 and passed to the multiplex system controller to be used as described below. Encoder rate control 1 also monitors encoder buffer fullness and uses this information to control the bit-rate of its encoder. Coded packets containing a variable number of bits send from the encoder to the encoder buffer.

Periodically, according to a predetermined system timing to be described, packets from the various streams are gathered together to form packs. Under control of the multiplex system controller, the multiplexer switch passes the so designated packets from the various encoder buffers to the multiplex buffer while they await transmission to the channel. The transfer of packets from the encoder buffers to the multiplex buffer is assumed to require only a fraction of a pack duration, so that a subsequent pack can be coded without undue risk of encoder buffer overflow.

System timing is maintained by the system clock. It is used in ways to be described and may also be inserted into the transmitted bit stream, for example, in the pack header data to enable the demultiplexing system to track accurately.

The operation of the de-multiplexing system is fairly simple. In the system of Fig. 8.5, incoming packets from the channel are identified as to which stream they belong to by the de-multiplexing controller, after which they are passed to the decoder buffers where they await decoding by the decoders. Each decoder waits a certain period of time after the arrival of the first bit of information from the channel before starting to decode. This delay is necessary to ensure that for any given access unit, the decoder has received all the bits for that access unit by the time that access unit needs to be displayed. Otherwise, decoder buffer underflow will occur.

Timing information is extracted by the de-multiplexing controller and fed to the system clock, which generates clock signal. Decoding and presentation timing information may also be included in the individual data streams by the encoders, to be used later by the decoders for synchronization of audio, video and other data [8-6]. In the absence of such timing information, satisfactory performance can often result if each decoder waits for some fixed delay LT after the arrival of the from bit of information from the channel before starting to decode.

In the system of Fig. 8.6 incoming packets from the channel are identified as to which stream they belong to by the packet selectors, after which they are passed to the decoder buffers where they await decoding by the decoders. In this system, system timing is passed to all decoders, which all keep their own independent time clocks.

In any real implementation, the decoder buffers will be of finite size. It is the responsibility of the multiplexing system to make sure that the decoder buffer do not overflow or underflow. In particular, each individual encoder rate controllers must guarantee that its encoder buffer does not overflow and its decoder buffer neither overflows nor underflows. Furthermore, the multiplex rate controller must guarantee that the combination of the encoder buffers and the multiplex buffer do not overflow and that no decoder buffer underflows. We now describe how this should be accomplished.

Statistical Multiplexing Algorithm: The statistical multiplexing algorithm adjusts the quantization to alter the video buffer input rate, and modifies the buffer output bit rate in order to optimize shared use of a fixed bandwidth by several video service encoders. In the implementation of such bit rate control, the following factors and goals must be considered:

1. A constant video signal quality (e.g. SNR) should be maintained over all types of frames (I, P, and B).
2. The MPEG-2 syntax sent with pictures that were processed in statistical multiplexing mode will indicate that the video stream is variable bit rate. Specifically, variable bit rate operation is defined in the bit_rate field sent in the sequence layer and the vbv_delay sent in the picture layer.
3. A bit rate change may only be implemented by a member of a statistical group when it is transporting a video packet. This initial video transport packet at the new bit rate must carry a PCR in its adaptation field.
4. The selected implementation must comply with the MPEG-2 Video Buffer Verifier (VBV) model.
5. The decoder video buffer should never underflow or overflow.
6. The encoder video buffer should never overflow.

Implementation of the statistical multiplexing algorithm usually require the following information periodically in order to adjust the quantization level:

1. The range of acceptable bit rates based on the encoder and decoder video buffer levels.
2. The encoder video buffer level for bit rate allocation and the quantization level determination.
3. Current picture status, including film mode, picture rate and picture type (e.g. I-frame or non-I-frame).

Both MPEG-2 Test Model and MPEG-4 Verification Model rate-control algorithms, discussed in Chapter 3, can be extended to the statistic multiplexing rate-control algorithm.

Next, we use the multiplexing system described in Fig. 8.4 to illustrate the basic concepts of statistical multiplexing algorithm.

In Fig. 8.4, each video-compression encoder generates a bit-stream and sends it to the corresponding encoder buffer. The multiplexer combines the output from all of the encoder buffers to form a final transport multiplexed stream. Statistical multiplexing algorithm is operated on the following stream group. The variable bandwidth individual video streams are grouped with other video streams to form a statistical group. The total bandwidth allocated to this group is fixed.

The quantization levels (QL) control the input bit-rate of each encoder buffer. The multiplex system controller controls the output packet rate of each encoder buffer. In statistical multiplexing both the input rate and output rate are adjusted in order to maintain a fixed bit-rate over multiple video services.

The idea behind statistical multiplexing is that individual video services in the group do not control their local QL themselves. Instead, the multiplex system controller provides a global QL for all the video elementary streams, and the local rate-control can only modify this QL if system robustness targets are not being met. As the complexity of each sequence varies and different picture types are processed, each encoder buffer fullness changes. The bit-rate assigned to each service by the multiplex system controller varies with this buffer fullness. In statistical multiplexing, the QL is more or less constant over the multiplex and the bit-rate changes reflect the sequence complexity changes. This ensures the more complex parts of a statistical group at a given time to be assigned more bandwidth, causing bandwidth to be used more efficiently over the entire multiplex. Note that this is also different with the fixed-rate operation, where the bit-rate of a video stream is fixed and the QL is changed to maintain this bit-rate.

The global QL value is computed based on the fullness of all the encoder buffers. Usually, the algorithm of generating the global QL needs to take different picture types into consideration. For example, consider an algorithm that is similar to the MPEG-2 Test Model rate-control algorithm described in Chapter 3. If only one virtual buffer is used for all picture types, in order to keep the buffer uniform over different pictures, corrections have to be applied based on the difference in picture sizes.

Bit rates for all video compression encoders are computed based on the fullness of each encoder buffer and buffer integrity (over- and/or underflow) checks. The bit-rate and QL for all the services are determined by means of exchanging information between the rate-control functions and the multiplex system controller.

One important feature for statistic multiplexing is to schedule I-pictures for each video encoders. This is, sometimes, also called I-picture refresh scheduling. In all video compression algorithms, I-pictures are the most important picture coding type. To ensure good video quality, more bits are usually spent on coding I-pictures. However, for a statistical multiplexing group, if I-pictures for each service are transmitted at the same time, QLs have to be increased for each video encoder. This will result poor compression quality. Hence, it is a task of the multiplex system controller to stagger I-picture generation of each video encoder so that the minimum number of video encoders which are members of the same statistical group will be outputting I-pictures at any given time. One simple method is to schedule I-picture refresh for a given statistical group by requesting I-pictures from each member in a round robin fashion.

Bibliography

For books and articles devoted to transport packet scheduling and multiplexing systems :

[8-1] ISO/IEC 13818-1:1996, Information technology – Generic coding of moving pictures and associated audio information: System, MPEG-2 International Standard, Apr. 1996.

[8-2] ITU-T Recommendation H.262 | ISO/IEC 13818-2: 1995. Information technology – Generic coding of moving pictures and associated audio information: Video.

[8-3] Test model editing committee, Test Model 5, MPEG93/457, ISO/IEC JTC1/SC29/WG11, April 1993.

[8-4] P. V. Rangan, S. S. Kumar, and S. Rajan, "Continuity and synchronization in MPEG", IEEE Journal on Selected Areas in Communications, Vol. 14, No. 1, Jan. 1996.

[8-5] D. K. Pibush, "Timing and Synchronization Using MPEG-2 Transport Streams," SMPTE Journal, pp.395-400, July 1996.

[8-6] Jae-Gon Kim and J. Kim, "Design of a jitter-free transport stream multiplexer for DTV/HDTV simulcast", Proceedings, JCSPAT'96, Boston, USA, pp. 122-126, Oct.1996.

[8-7] J. G. Kim, H. Lee, J. Kim, and J. H. Jeong, "Design and implementation of an MPEG-2 transport stream multiplexer for HDTV satellite broadcasting", IEEE Transactions on Consumer Electronics, vol. 44, no. 3, August 1998.

[8-8] Xuemin Chen, "Rate control for stereoscopic digital video encoding", US Patent Number 6072831, Assignee: General Instrument Corporation, June 6, 2000.

[8-9] B. G. Haskell and A. R. Reibman, "Multiplexing of variable rate encoded streams", IEEE Trans. on Circuits and Systems for video technology, Vol. 4, No.4, August 1994.

[8-10] Xuemin Chen, Fan Lin, and Ajay Luthra, "Video rate-buffer management scheme for MPEG transcoder", WO0046997, 2000.

[8-11] Xuemin Chen and Fan Ling, "Implementation architectures of a multi-channel MPEG-2 video transcoder using multiple programmable processors", US Patent No. 6275536B1, Aug. 14, 2001.

[8-12] B. G. Haskell, A. Puri, and A. N. Netravali, Digital Video: An Introduction to MPEG-2, New York: Chapman & Hall, 1997.

[8-13] A54, Guide to the use of the ATSC digital television standard, Advanced Television Systems Committee, Oct. 19, 1995.

[8-14] A. R. Reibman and B. G. Haskell, "Constraints on variable bit-rate video for ATM networks", IEEE Trans. On Circuits and Systems for video technology, Vol. 2, No.4, Dec. 1992.

[8-15] Jerry Whitaker, DTV Handbook, 3rd Edition, McGraw-Hill, New York, 2001.

9 Examples of Video Transport Multiplexer

Two examples of video transport multiplexer are introduced in this chapter to illustrate many design and implementation issues. One example is an MPEG-2 transport stream multiplexer in encoder and other is an MPEG-2 transport re-multiplexer. As discussed in the previous chapters, MPEG-2 consists of a coding layer around which is wrapped a system layer [9-1]. Whereas the coding layer handles compression and decompression [9-2][9-3], the system layer handles streaming, continuity, and synchronization. The system layer packetizes compressed media units, interleaves the packets of different media and organizes them into a transport stream. The TS packet is the basic unit for maintaining continuity of decoding, via a system clock reference (PCR) time stamp that is inserted in the adaptation layer of TS header. The PES packet is the basic unit for maintaining synchronization between media playback, via decoding and presentation time stamps (DTS and PTS) inserted in the packet headers. Whereas MPEG provides the framework for insertion, transmission and extraction of these time stamps, additional feedback protocols are used to make use of these time stamps in the enforcement of synchronization, particularly in a distributed multimedia environment.

9.1 An MPEG-2 Transport Stream Multiplexer

An example of design and implementation of MPEG-2 transport stream multiplexer is provided in this section. Such example is introduced only for educational proposes in order to illustrate design principles, implementation considerations and architecture of MPEG-2 transport stream multiplexer [9-4][9-6][9-7].

As introduced in Chapters 1, 5 and 8, the MPEG-2 system standard [9-1] has been widely used as a transport system to deliver compressed video and audio data and their control signals for various applications such as digital television broadcasting. As illustrated in Section 1.3, MPEG-2 systems specification provides two methods for multiplexing elementary streams (ES) into a single stream. MPEG-2 systems specification also provides a function of timing and synchronization of compressed bit streams using timing elements. MPEG-2 transport stream (TS) is primarily used for error-prone environments, such as satellite and cable transmission.

The digital video multiplexer discussed in this section is an MFEG-2 TS multiplexer with special considerations for timing and synchronization. In particular, a scheduling algorithm is described which uses information of the buffers of T-STD (reviews in Chapter 8 in details) as one of scheduling factors.

9.1.1 Overview of the Program Multiplexer

The program multiplexer of MPEG-2 TS discussed here [9-7] combines elementary streams of one video and two audio signals into a single MFEG-2 TS, which ensure timing and synchronization constraints of the standard.

As shown in Fig. 9.1, one video and two audio elementary streams output from video and audio encoders are sent to the program multiplexer. Before multiplexing, packetized elementary stream (PES) packets are generated for video and audio data by the program multiplexer. Then, a stream of TS packets (of length 188 bytes) is generated by multiplexing the PES packets with additional packets including the program specific information (PSI).

As discussed in Chapter 1, synchronization of decoding and presentation process for audio and video at a receiver is a particularly important aspect of a real time program multiplexer. Loss of synchronization could lead to either buffer overflow or underflow at the decoder, and as a consequence, loss of presentation synchronization. In order to prevent this problem and ensure precise presentation timing, MPEG-2 System specifies a timing model in which the end-to-end delay through the entire transmission system is constant [9-1][9-5]. This is achieved by generating the transport stream with two types of timing elements: program clock reference (PCR) and presentation/decoding time stamp (PTS/DTS).

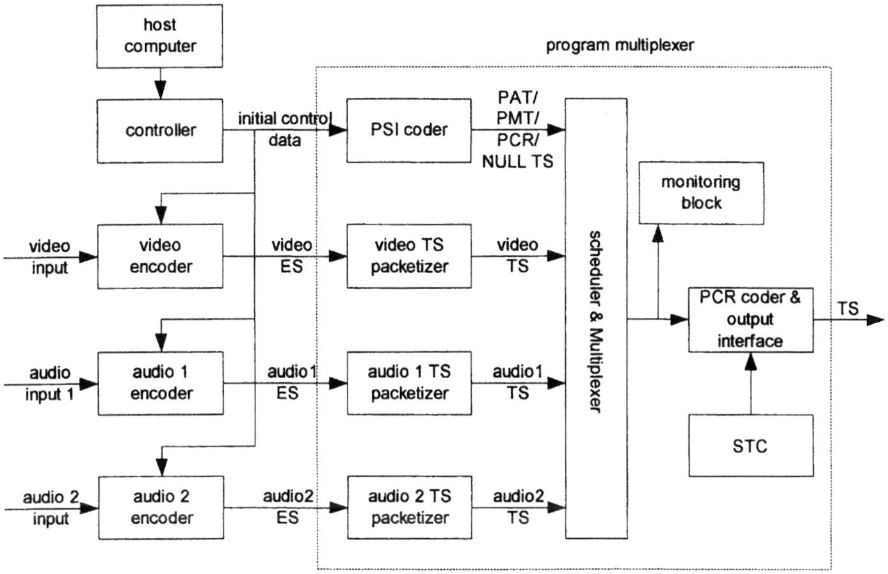

Figure 9.1 Block diagram of a program multiplexer.

PCR is a sampled value of the encoder's 27 MHz system time clock (STC) and is periodically inserted by a PCR coder in the adaptation headers of a particular TS packet named PCR packet. PCR serves as a reference for system clock recovery at the decoder and establishes a common time base through the entire system. Synchronization between video and audio signals is accomplished by comparing both the audio presentation time stamps (PTS) and the video PTS with the STC.

These timing elements are defined precisely in terms of an idealized hypothetical decoder: the transport stream system target decoder (T-STD) which is also used to model the decoding process for exactly synchronized decoding and presentation. Therefore, the transport stream generated by the program multiplexer should comply with the specifications imposed by T-STD model to achieve normal operations of real time decoding process. The precisely coded time stamps PCR and DTS/PTS are embedded in the output Transport stream.

In particular, the monitoring block observes the behavior of buffers of T-STD. A scheduler that determines the order of TS packets uses the information obtained from the monitor block as key control parameters to ensure the restrictions imposed by T-STD being satisfied. To provide flexibility to multiplexing mode, the host computer with a system controller is applied to the program multiplexer.

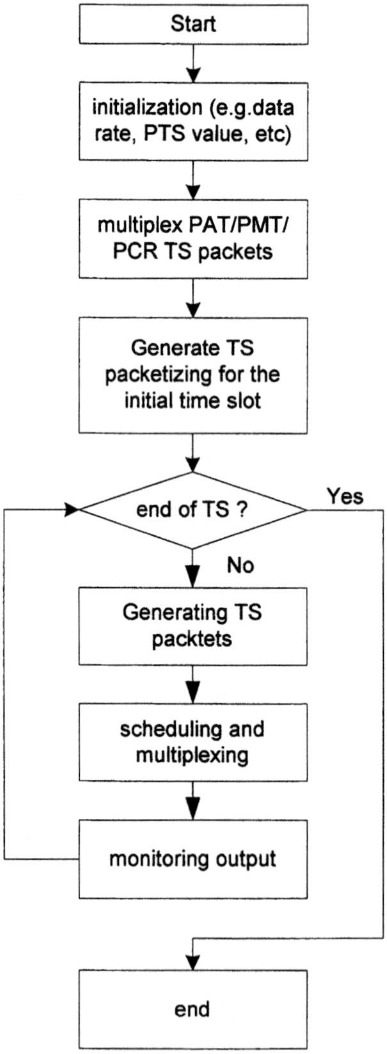

Figure 9.2 Software flow-diagram for TS generation

8.4.2 Software Process for Generating TS Packets

First, consider a software TS packet generating process. In order to emulate a real time hardware operation, lets set the observing time slot T_{TS} as a time needed for one TS packet to be transmitted at a given transport rate R_{TS} (bits/second). Since the length of a TS packet is 188 bytes, T_{TS} is obtained as follows:

$$T_{TS} = \frac{188 \cdot 8}{R_{TS}}. \tag{9.1}$$

The software TS packet generator emulates hardware function blocks at every time slot. As shown in Fig. 9.2, this generator comprises several detailed blocks that are directly mapped to those blocks in Fig. 9.1.

In the TS generation process, each iteration loop includes following steps.

1. Initialization of parameters: data rates of each ES and TS, the total number of TS packets to be generated, transmission rate of PCR and PSI, observing time slot T_{TS}, and the value of initial PTS.
2. Multiplex three types of packets: PAT, PMT, and PCR packets at the beginning of the process for two time slots.
3. TS packet generation, scheduling and multiplexing, and output monitoring. Storing TS packets in an output buffer in the muitiplexing block.
4. Repeat the step 3 until the end of the stream.

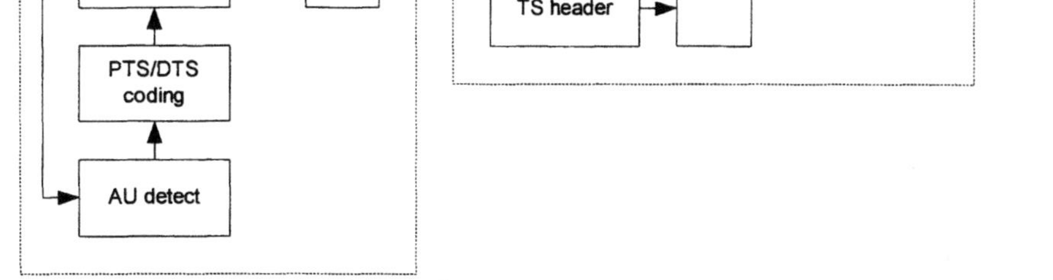

Figure 9.3 A block diagram of TS packet generation

TS Packet Generation: Elementary streams are packetized in the TS generation block as shown Fig. 9.3. It usually consists of a PES packetizer and a TS packetizer. The output TS packets are stored in the main buffer and these packets will be multiplexed into a transport stream.

In each time slot, the PES packetizer fetches a fixed amount of ES data from encoder output buffer. The ES bytes n^j fetched in the j-th time slot are set to be an integer and the remainder bits r^j will be sent in the next time slot as given below,

$$d^j = T_{TS} \cdot R_B + r^{j-1}, j=1,2,\ldots\ldots \qquad (9.2)$$

$$r^0 = 0,$$

$$r^j = d^j - n^j,$$

$$n^j = \left\lfloor \frac{d^j}{8} \right\rfloor,$$

where R_B denotes the ES bit rate in bytes per second, and $\lfloor \bullet \rfloor$ denotes the floor operator.

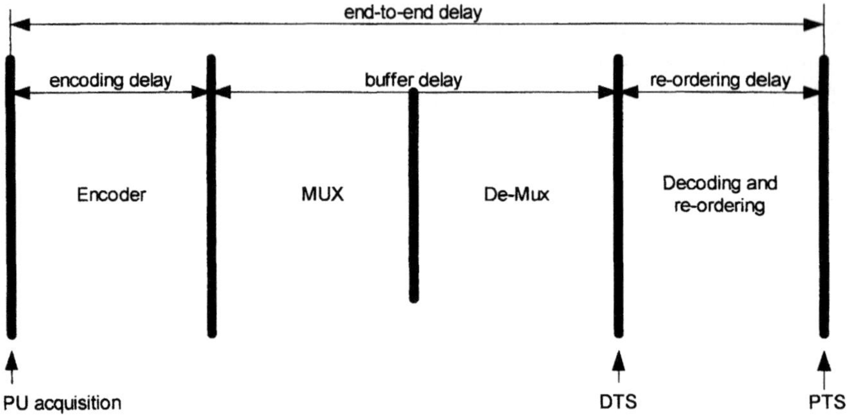

Figure 9.4 The end-to-end system timing model

PES packetizer detects access units (AU's) of each ES for inserting PTS/DTS into PES header. Since video AU's (coded pictures) have variable sizes and MPEG-2 layer-1 and layer-2 audio or AC-3 audio AU's have a fixed size, PES packetizer produces variable-length PES packets for video and fixed-length PES packets for audio. PES packet data are then stored in the buffer 2 until they are loaded into TS payload.

A TS packet is generated at each time slot when there is at least one payload of 184 bytes are stored in buffer 2. PES packets are aligned to TS packets with the PES header being the first bytes in the TS payload. This alignment is accomplished by placing stuffing bytes in the adaptation field of the packet preceding the PES aligned packet. Therefore, the payload of the TS packet may have length of 1~184 bytes to achieve the alignment.

Time stamping: As aforementioned, constant end-to-end delay is maintained in the timing model by using the PCR and PTS/DTS time stamps as shown in Fig. 9.4.

Based on MPEG-2 systems specification, PCRs are transmitted periodically at least once every 0.1 second by using PCR packets. If a PCR packet is the *N*-th packet, the value of PCR that coded in the field of *pcr_base* and *pcr_extension*, are calculated as

$$m = N \cdot T_{TS},$$
$$pcr\_base(m) = \phi(m),$$
$$pcr\_extension(m) = \psi(m),$$

where

$$\phi(m) = \left(\left(system\_clock\_frequency \cdot m \right) / 300 \right) \% 2^{33},$$
$$\psi(m) = \left(system\_clock\_frequency \cdot m \right) \% 300.$$

Thus, PTS for each AU can be calculated as

$$PTS(i) = \phi(T_{acq}(i)) + \phi(D_{sys}), \; i = 1,2,... \tag{9.3}$$

where $T_{acq}(i)$ denotes the acquisition time of the *i-th* presentation unit and D_{sys} denotes the system end-to-end delay. That is, the PTS for the *i-th* AU, $PTS(i)$, is coded as sum of the value of STC at the AU acquisition time at the encoder and the value of STC according to the end-to-end delay. The synchronization of presentation between video and audio is achieved by setting the first PTS of them to be the same value.

The end-to-end delay consists of encoding delay, buffer delay of STD and video reordering delay as shown in Figure 9.4. In this model, it is assumed that there is no encoding delay because the ES of audio and video are stored in files and the first presentation unit (PU) is acquired as soon as the generator start-up. Then PTS are given by

$$PTS(1) = \phi(D_{sys} - D_e),$$
$$PTS(i) = PTS(i-1) + \phi(T), i = 2,3,... \tag{9.4}$$

where D_e denotes the encoder delay and T denotes the period of PU, e.g. the picture duration of the original uncompressed video and $\phi(T)$ denotes the nominal frame time in 90 KHz clock cycles (see Chapter 5 for details), e.g. it equals 3003 in NTSC video where $T = \frac{1}{29.97}$ while it equals 2160 in AC-3 audio at the sampling rate of 48kHz.

For the MPEG-2 compressed NTSC video that requires picture reordering, the DTS is updated by the following equations according to the MPEG-2 coding structure - Group Of Pictures (GOP):

$$DTS(1) = PTS(1) - 3003 \text{, for 1st I-picture.}$$

$$DTS(i) = PTS(i) - 3003 \cdot M \text{, for P-pictures and other I-pictures,}$$

$$DTS(i) = PTS(i) \text{, for B-pictures,}$$

where M is distance between P-pictures in a GOP.

Scheduling and Multiplexing: The scheduler determines the types of TS packets being sent at a given time slot and the multiplexing block produces a constant bit rate of transport streams. For the discussed system, TS packets contain a video, two audio, PSI and null packet data. Normally, the output TS data rate of program multiplexer is greater than sum of the combined data rates of all ES's and the data rate of systems layer overheads. In this case, null packets are inserted to keep the constant bit rate of TS. Two important types of packets to carry PSI data are program association table (PAT) and program map table (PMT) packets.

The scheduling algorithm needs to consider the following parameters:
1. Transmission priority for each packet type,
2. Fullness of the main buffers for video and audio,
3. Transmission rate for PCR and PSI,
4. Output monitoring results for validation of the T-STD.

Usually, the priority of the packet types is ranked in the order of PAT, PMT, and PCR packets in the initial state. In normal state, the packets have priority in the order of PCR, PAT, and PMT packets. For the packets that contain ES, audio packets often have a higher priority than video. The fullness of the main buffers represents the amount of PBS data of video and audio waiting for TS multiplexing.

Assume that a PCR packet is sent at least every T_{PCR}. The transmission period of PCR packet in terms of TS packet number can be set as

$$N_{PCR} = \left\lfloor \frac{T_{PCR}}{T_{TS}} \right\rfloor . \tag{9.5}$$

The scheduling block would simply count packets until it reaches N_{PCR} and then sends a PCR packet. The period for sending PSI packets can be scheduled in the same way. The generated TS must guarantee that the operations of T-STD follow the TS specification. This is accomplished by considering the output observing result when it comes to scheduling at every time slot.

8.4.3 Implementation Architecture

In this subsection, software process of generating TS packets is directly mapped to the hardware implementation architecture.

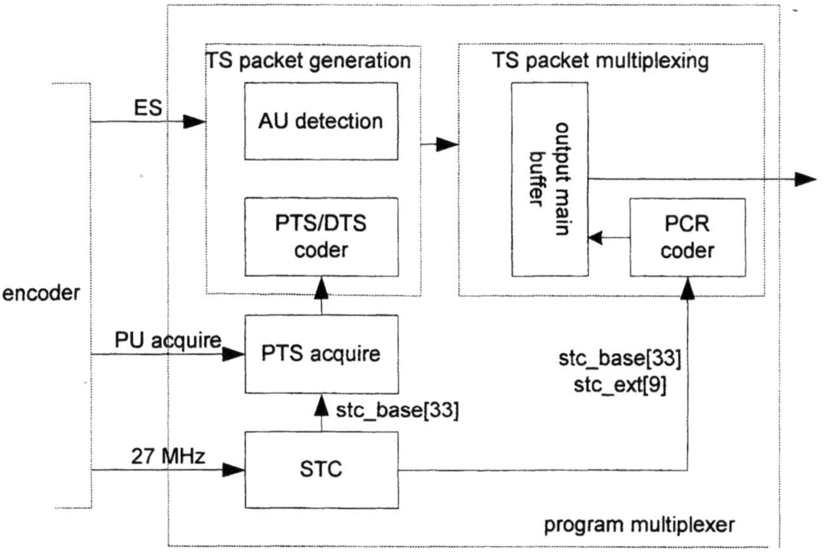

Figure 9.5 Coding of time stamps

Time-stamping: Fig. 9.5 shows the time stamping process of the program multiplexer. In order to encode PTS, the value of the STC is sampled first whenever a PU acquisition signal (PU acquire) is activated from the video encoder. Then, the PTS/DTS coder inserts the sampled value into PTS field when the corresponding AU is detected from incoming ES. The PCR value is inserted in the last stage as shown in Fig. 9.1 to maintain constant delay between the instant where the PCR is inserted to the stream and decoding of that field [9-6]. The PCR coder detect the PCR packet from multiplexed transport stream and exchange the dummy PCR value with the time value that is sampled at the instant of PCR packet transmission.

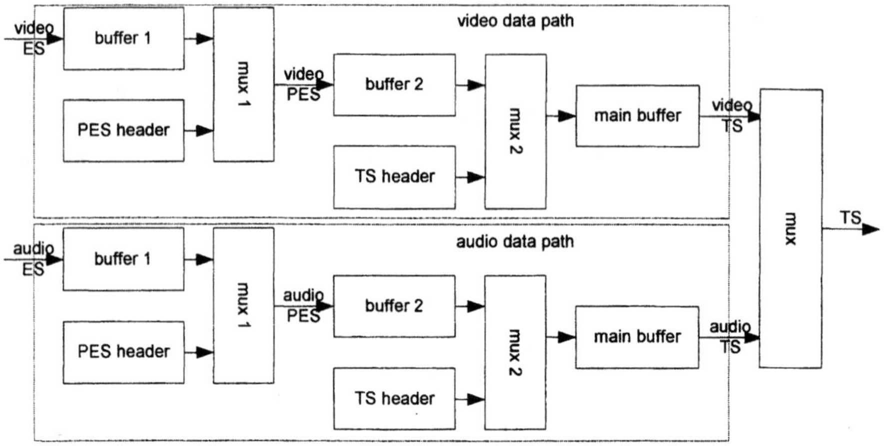

Figure 9.6 Data buffers for TS generation

Data Flow and Buffering: Fig. 9.6 shows data flow and buffering for video and audio paths in the program multiplexer. It is necessary to include buffering in the process of PES and TS packetizing and packet-based multiplexing.

Both buffers 1 and 2 perform the functions of PES and TS packet overhead buffering specified in the T-STD model. These buffers are implemented with FIFO memory and have the size within the bound of overhead buffering. The main buffer that accommodates the delay caused by packet multiplexing is mapped for additional multiplexing buffer B_{mux} of T-STD with the size of BS_{mux} (see Chapter 8 or reference [9-1] for definitions).

The buffer 1 in the audio path also includes a time delay buffer to compensate for the difference of encoding delay between video and audio encoders. It is essential to prevent all buffers to overflow or underflow. The scheduler plays the key role to maintain the buffers in normal operation.

Scheduling and Multiplexing: Fig. 9.7 shows the block diagram for scheduling and multiplexing. As mentioned before, the scheduler determines which packet type should be sent at a given time slot based on proper conditions. The main buffers of video, audio 1 and audio 2 activate control signals of v_ready, *a*1_ready and *a*2_ready when more than two TS packets are buffered, respectively. Similarly, the interrupt generates pcr_request and psi_request signals in their pre-determined transmission intervals.

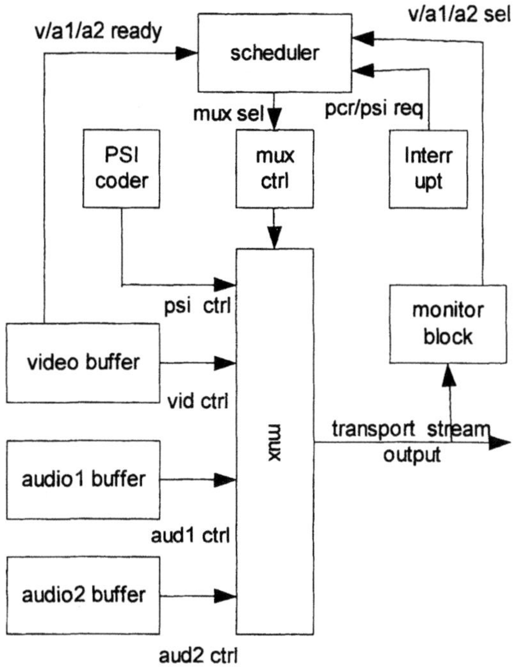

Figure 9.7 Block diagram of TS packet multiplexing

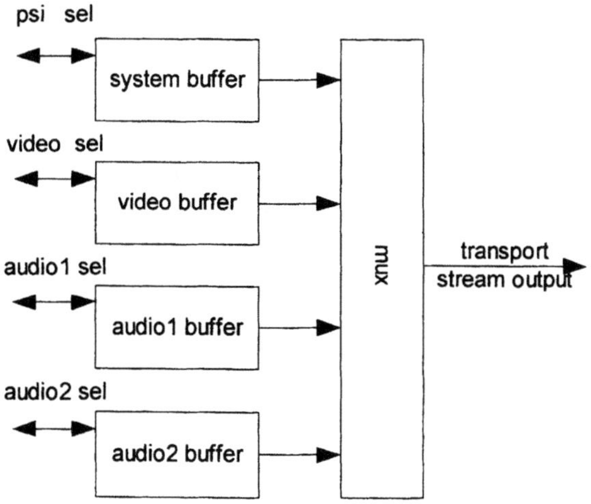

Figure 9.8 Block diagram of buffer monitoring and selection

Also, the monitor block observes the status of buffers and generates select signals that indicate the specific packet multiplexed for the next time slot.

This block consists of separated monitoring sub-blocks for video, audio and system data as Fig. 9.8. Each sub-block includes several hypothetical buffers specified in T-STD, implemented by a simple up-down counter logic to check the fullness of corresponding buffers.

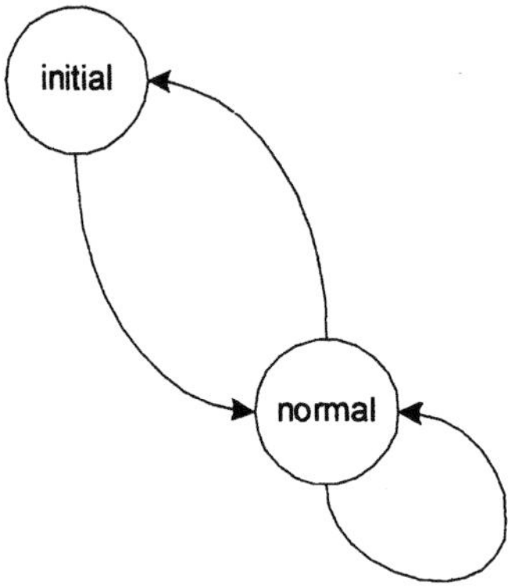

Figure 9.9 The state machine for the scheduler

The scheduler is implemented by using a state machine as shown in Figs 9.9 and 9.10. The output signals represent the selected packet type for the time slot. At the beginning of TS generation, three packets in order of PAT, PMT, PCR packets are selected in the initial state. In normal state, all of the packets are scheduled according to described control signals.

The host-computer and controller provide user interface for operational mode setting, start-up timing control and running state monitoring for each encoder modules including the program multiplexer as shown in Fig. 9.1. The controller software is downloaded from the host-computer and is responsible for generating the data required for initialization of operation modes. These include the PSI contained in the PAT and PMT, PID's setting, ES data rates for video and audio and TS rate and transmission periods of PSI and PCR, etc..

The function blocks described here can be implemented in DSP, macro-controller, or other devices.

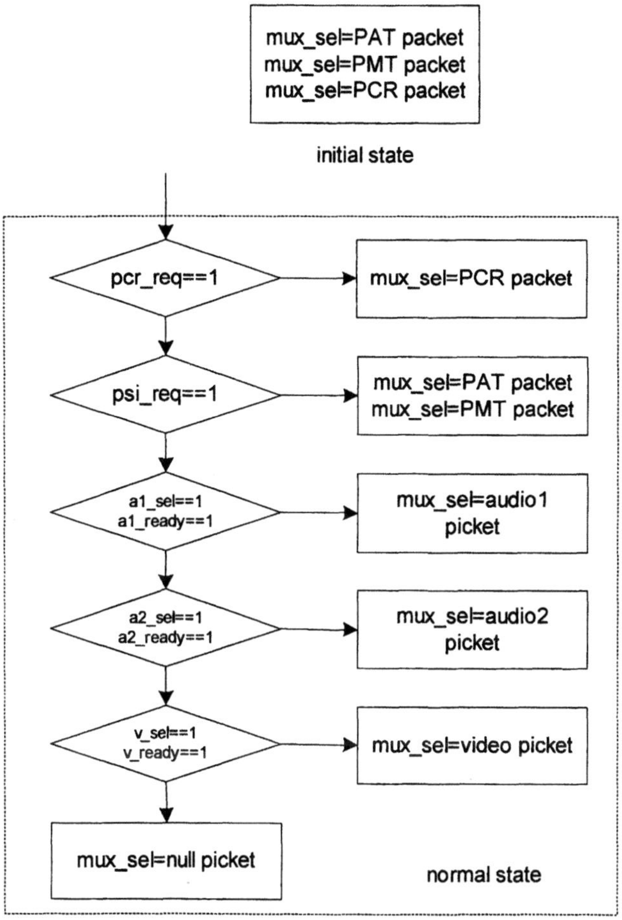

Figure 9.10 Control signal flows for initial and normal states

9.2 An MPEG-2 Re-multiplexer

A re-multiplexer (or simply called "ReMux") is a device that receives one or more MPEG-2 multi-program transport streams (TSs) and retains a subset of the input programs, and outputs the retained programs in such a manner that the MPEG timing and buffer constraints on output streams are satisfied. A video transcoder can be used along with ReMux to allow bit-rate reduction of the compressed video [9-8][9-9][9-10]. Again, the example introduced here is only for educational proposes.

In digital television services, a ReMux is usually used in parallel with other video, audio, and data encoders, all of which feed into a common output multiplexer as shown in Fig. 9.11.

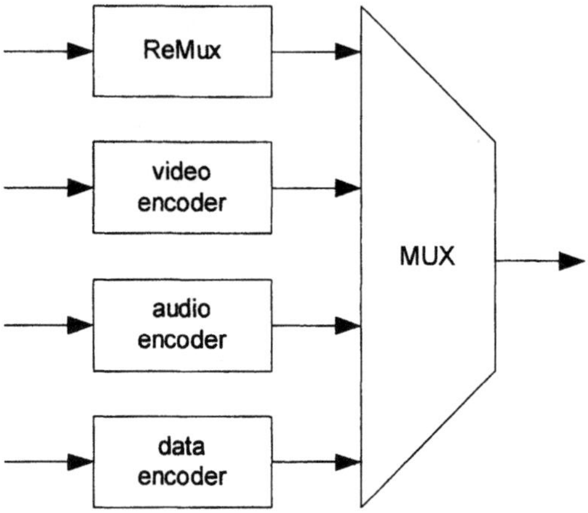

Figure 9.11 A DTV Multiplex System with ReMux

The ReMux can enable many new services over digital video networks. For example, the ReMux enables a television service provider to combine into a single bitstream remotely compressed bitstreams and/or precompressed stored bitstreams with locally material.

In general, the ReMux can operate on both constant bit rate (CBR) and variable bit rate (VBR) modes as defined and described in the previous chapters. In the CBR mode, the ReMux is often configured during initialization with the bit-rate of the traffic it is to retain and output. To provide a better quality of service, many services would prefer to use a ReMux that efficiently support VBR bitstreams. In this section, we illustrate the design principles of the ReMux by using a VBR example.

9.2.1 ReMux System Requirements

In Chapter 5, we discussed the MPEG-2 system multiplexer. Similar to the multiplexer, the main function of a ReMux is to schedule the output of packets from all of its input packets. The ReMux performs this function through a packet scheduler that generates a list of the order for outputting packets. The packet scheduler supports applications with different demands on bandwidth. It is constructed in such a way that a smooth output is generated with relatively equal spacing between the packets for any individual application.

First, let us briefly review some of the fundamentals of MPEG-2 Transport Streams (TSs). As being described in Chapter 8, the MPEG-2 TS contains two types of timestamps:

- Program clock references (PCRs) are samples of an accurate bitstream-source clock. A MPEG decoder feeds PCRs to a phase-locked loop to recover an accurate timebase synchronized with the bitstream source.
- Decoding time stamps (DTSs) and presentation time stamps (PTSs) tell a decoder when to decode and when to present (display) compressed video pictures and audio frames.

MPEG-2 Systems standard specifies a decoder behavioral model [9-1] and all compliant TSs can be successfully decoded by such model. The model consists of a small front-end buffer called the transport buffer (TB) that receives the TS packets for the video or audio stream from a specific program identifier (PID) and that outputs the received TS packets at a specified rate. The output stream of a TB is sent to the decoder main buffer(s), denoted by B. B is drained at times specified by the DTSs. A simplified diagram of MPEG-2 Systems decoder model is shown in Figure 9.12.

All legal MPEG-2 encoders produce bitstreams that are decodable successfully by this model:

- bit-rates and TS packet spacing are appropriate to ensure TB does not overflow.
- DTSs/PTSs ensure that video and audio frames can be decoded and presented continuously, without overlaps or gaps (e.g. 29.97Hz for NTSC video).
- DTSs/PTSs and coded frame sizes are determined in by the encoder such that B neither overflows nor underflows.

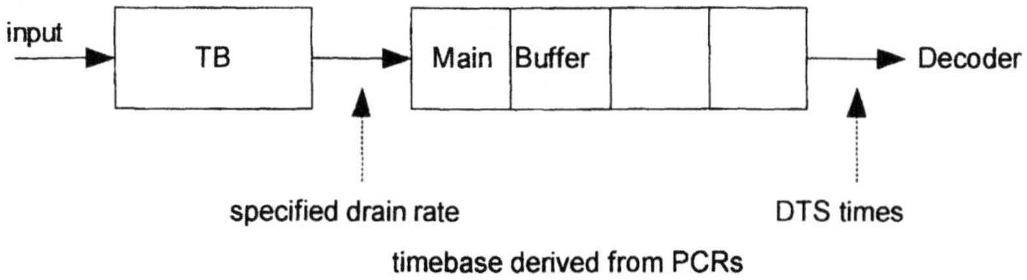

Figure 9.12 A simplified diagram of MPEG-2 Systems decoder model.

The challenge for a ReMux design is to accept a legal TS as input, to discard certain components from the input TS, and to manage the multiplexing of the retained traffic with locally encoded traffic such that the resulting output bitstream also complies with the MPEG model. This is difficult because constraints on the ReMux, and the fact that packets from different applications can become available at the same time, force the ReMux to delay different packets by different amounts.

For VBR applications, the ReMux can change its packet schedule in each schedule period. During each schedule period, the ReMux will

- Collect activity information from each VBR video application,
- Assign bit-rates to the VBR video applications,
- Communicate the bit-rates to the VBR video applications,
- Create a packet schedule that reflects the VBR video rates (and the rates of CBR applications) for the schedule period.

In real implementation, the ReMux does not actually schedule the new packet until time *T*, where *T* is the look-ahead interval. In other words, the ReMux does not assign a bit-rate to the data segment in its schedule period. Instead, it calculates the bit-rate for a data segment by using data buffered in the look-ahead interval, i.e. the ReMux buffers data for more than the look-ahead interval. This look-ahead time is also needed to provide the best possible quality for video compression.

9.2.2 Basic Functions of the ReMux

Basic functions of the ReMux include:

- Smoothing the possible burst input traffic,
- Discarding the programs from the original TS stream that are not supported by the service,
- Estimating the rate of the retained traffic, in advance, for each schedule period,
- Determining the bit-rate for the data in a look-ahead manner to ensure the ReMux having sufficient bandwidth to output its retained traffic with best video quality.

A block diagram is given in Fig. 9.13 for a ReMux to perform the above functions.

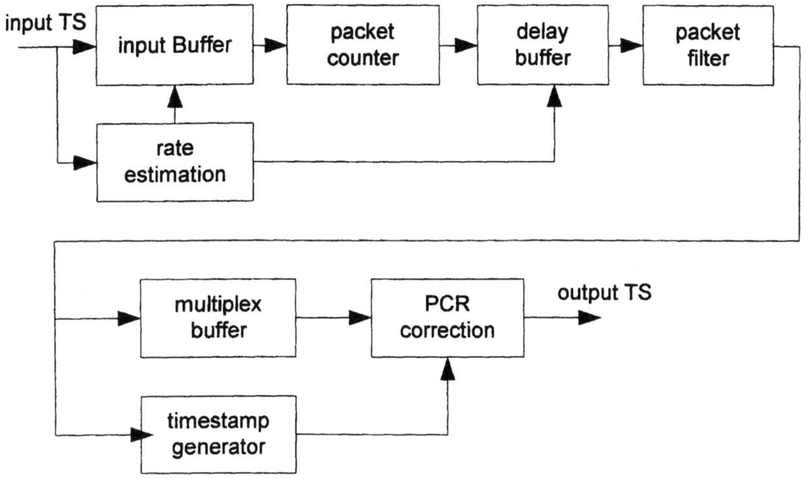

Figure 9.13 A functional diagram of a ReMux

A real ReMux implementation may need to perform additional functions such as providing an interface for selection of discarded and retained traffic, handling program system information data, supporting diagnostics, etc.). However, Fig. 9.13 provides a high-level implementation diagram of the key re-multiplexing function of a ReMux.

Assume that a multiprogramming TS is fed into the ReMux. Usually, MPEG-2 requires that the multiprogramming TS should be a CBR stream [9-1]. Thus the input bit-rate to the input buffer in Fig. 9.13 is constant. In practice, the actual input TSs' rates may be piecewise-constant. The constituent programs in a multiprogramming TS need not be CBR, but the sum of the constituent bit-rates must be constant.

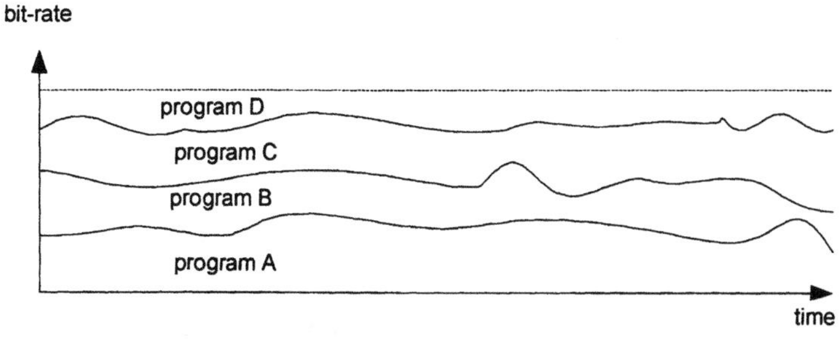

Figure 9.14 A CBR multi-program TS is made up of variable-rate constituents

The rate estimator in Fig. 9.13 estimates the input rate R_{in} of the ReMux. Such task is complicated by the fact that the input bitstream may not be delivered smoothly and continuously because of network behavior, but is simplified by the fact that the input rate, averaged over reasonably long times, is fixed. The input buffer stores the input bitstream and outputs a smooth and constant-rate bitstream at the rate provided by the rate estimator. The ReMux usually implements the rate estimator with control software that is given snapshots of the input buffer fullness in a given time-interval. The software assigns the output rate of the input buffer such that:

- The input buffer does not overflow or underflow, i.e. the long-term average input rate and output rate of the input buffer are equal.
- After initialization, the output rate of the input buffer changes sufficiently slowly that the MPEG-2 system clock frequency slew rate limitation is not violated (see [9-1] section 2.4.2.1). System clock frequency slew is created if different system timestamps (e.g. PCRs for MPEG-2 transport and SCRs for DirecTV transport) traverse through the ReMux with different delays.

The output of the input buffer is a nearly exact replica of the input bitstream as it was originally encoded, i.e. without transmission delay jitter.

The packet counter block in Fig. 9.13 performs two functions:

- It tags transport stream packets that are to be discarded, e.g. packet streams indicated by the ReMux user and the packet stream with the packet identifier value for NULL packets (PID=0x1FFF for MPEG-2 transport streams). This facilitates easy discard of these packets later.
- It counts the number of retained (i.e. not discarded) packets that arrive at the block. In each packet count interval T_1, the number P_j of retained packets passed through the queue is counted, shown in Fig. 9.15. In every scheduling period T_2 where $T_2 = nT_1$, software on the ReMux reads all of the packet counts that have been queued during the previous schedule period, and calculates the ReMux output rate R_{OUT} corresponding to this schedule period.

Packets output from the packet counter block enter the delay buffer. At the initialization time, the delay buffer depth is configured so that the delay of the delay buffer is $2 \cdot T_2 + T_3$ where T_3 denotes the ReMux look-ahead interval. (shown in Fig. 9.15). Then, once the rate estimator determines the bit-rate R_{in} of the incoming bitstream, the delay buffer depth is configured to (and fixed at) $(2 \cdot T_2 + T_3) \cdot R_{in}$ bits.

At the output of the delay buffer in Fig. 9.13 is the originally encoded bitstream, with its original timing (restored by the input buffer). Assume that this bitstream obeys all of MPEG's timing and buffer constraints, since it is a nearly-exact replica of a bitstream from an originating encoder.

The packet filter removes those earlier tagged packets for discard and output retained packets. If the ReMux could deliver all retained packets at exactly the same time as they occur in this version of the bitstream, one would be assured that all MPEG constraints for the retained streams would be obeyed. However, this is impossible because when some constituents of the original input stream are removed, e.g. program D in Fig. 9.14, the total bit-rate of remaining constituents is usually a variable. Thus, the ReMux must change the output timing of retained packets somewhat. At the output of the packet filter, retained PCR packets are detected by the ReMux timestamp generator for computing new PCR values. The process given in Chapter 7 for regenerating PCR can be used in here.

Next, all retained packets pass into the multiplex buffer. This buffer size is at least *N* bits deep, where N is the total multiplex buffer size of all retained elementary streams (see Chapter 8 or reference [9-1] for definitions). Packets are removed from the multiplex buffer when they are requested. If a packet being output contains a system timestamp, the system timestamp is incremented by the local timestamp generator value.

9.2.3 Buffer and Synchronization in ReMux

ReMux Output Rate: In every T_1 seconds, the number of retained packets is stored in a queue. In every T_2 seconds, the queued counts for the previous T_2 seconds are scanned to determine the ReMux output bit-rate R_{OUT} corresponding to the previous (T_2 second-long) ReMux scheduling period. The calculation is performed as follows:

1. Determine the highest number $P_{\max} = \{P_j : j = 1,2,\ldots,n\}$ of retained packets that arrived at the ReMux in n T_1 interval of a scheduling period T_2.
2. Calculate the output rate for the ReMux schedule period by using

$$R_{OUT} = \frac{1504 P_{\max} \cdot K_s}{T_1}, \tag{9.6}$$

where $K_s \geq 1$ is a scale factor that is determined by application and 1504 is the value of bit per packet. Fig. 9.15 shows the timing relation for such computation.

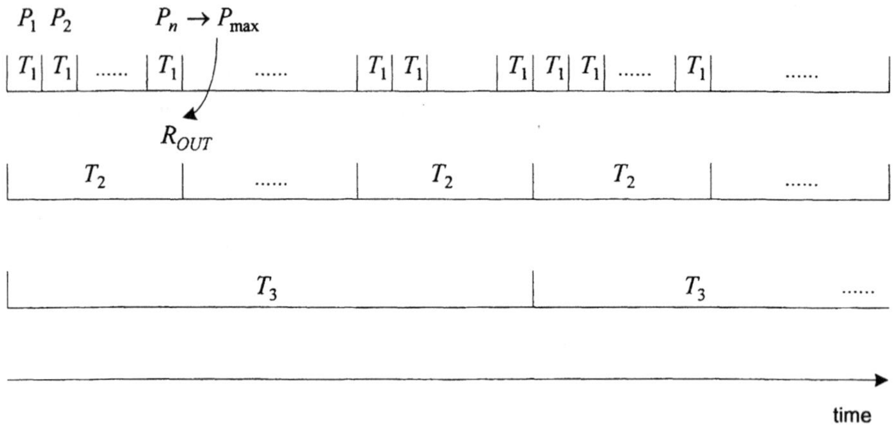

Figure 9.15 Timing relation for the ReMux

Synchronization: The ReMux is signaled at the start of each ReMux scheduling period T_2 by a broadcast message. After the broadcast message, T_2 packet times or seconds pass until a new ReMux schedule actually goes into effect. Since the ReMux look-ahead interval T_3 is an integer multiple of the ReMux scheduling period T_2, and $T_3 = mT_2$, the ReMux knows almost exactly which T_1-count values correspond to each ReMux scheduling period. (Inaccuracy may result from uncertainty in T_1 and T_2 times, delay buffer depth, etc.) During each ReMux scheduling period, the Remux calculates the rate needed for the previous ReMux schedule period's worth of data to enter the delay buffer.

Buffer Headroom: MPEG allows originating encoder to operate the main decoder buffer (B or MB+EB) at a nearly empty level. Thus, the ReMux cannot cause this buffer to become emptier. However for ReMux, MPEG reserves some "headroom" in the main buffer specifically to aid in re-multiplexing. The ReMux can cause the main buffer to run slightly fuller than in an original TS. This headroom, specified by MPEG, is different for video and audio bitstreams, but in all cases it can hold more than 4 msec worth of data [9-1]. The ReMux can use this headroom to limit its movement of packets.

The ReMux can control the fullness of the main buffer by varying PCR values while holding PTS/DTS values fixed. For example, if the ReMux makes PCR values smaller, then PTSs become larger with respect to their bitstream's time-base, so frames are decoded later and the main buffer is fuller.

PCR Correction: As described in section 9.2.2, the ReMux adjusts all retained PCR values. The ReMux adjusts retained streams' PCRs such that with no delay through the multiplex buffer, each retained stream has its decoder buffer somewhat fuller than before the adjustment. Since decoder buffers are to be fuller, PCR values are made smaller. The amount of the adjustment is

$$A = \min_{ES \in retained\ packets} \left\{ \frac{HS_{ES}}{R_{ES}} \right\}, \tag{9.7}$$

where HS_{ES} denotes the headroom size for the ES and R_{ES} denotes the rate of the ES. The A value is chosen such that there is at least one retained elementary stream whose headroom is made full because of the adjustment. In some implementation, this value might be calculated more simply or even might be fixed. When the multiplex buffer (in Figure 9.13) is not empty, the PCR can be adjusted by the value of (output buffer delay - A). In the case that the multiplex buffer is full, the multiplex buffer delay is A and the PCR adjustment is 0.

MPEG Buffer Verification: The ReMux multiplex buffer can underflow in normal operation without any problems. In fact, if the Mux in Fig. 9.11 serves the ReMux at a rate much higher than the rate of the retained traffic, then the ReMux multiplex buffer is always nearly empty, and the delay through the ReMux multiplex buffer always is nearly 0. In this case, the PCR adjustment is needed to ensure that each re-multiplexed bitstream consumes extra decoder buffer space, but less space than allowed by the MPEG headroom.

Note that If the ReMux multiplex buffer delay is more than A, then the ReMux has delayed packets sufficiently to cause decoder main buffer to be emptier than they should be. This might cause decoder buffer underflow. It has to be ensured that for each packet that it processes, the ReMux keeps the packet's delay through its multiplex buffer less than A. The ReMux does this heuristically: the choice of scale factor K_s given in Eq.(9.6), should be a carefully selected value such that the estimated rate of each ReMux schedule period can keep the ReMux multiplex buffer empty enough to satisfy this constraint.

Monitoring Transport Buffer: A remaining problem with the above algorithm is that it may change the output timing of retained packets in such a way that the packets could cause TB overflows. Simply increasing the ReMux output rate for the current scheduling interval often can not solve the problem--it outputs packets earlier than if the rate were lower and thus makes the TB overflow problem more severe. (one would want to increase the output rate of previous scheduling intervals.) Again, the ReMux can selects a proper scale factor K_s to solve this problem heuristically: when the Mux in Fig. 9.11 serves the ReMux at a rate slightly higher than truly needed, the ReMux multiplex buffer stays near empty, which keeps packets' output times close to their original output times.

Bibliography

[9-1] ISO/IEC 13818-1:1996, Information technology – Generic coding of moving pictures and associated audio information: System, MPEG-2 International Standard, Apr. 1996.

[9-2] ITU-T Recommendation H.262 | ISO/IEC 13818-2: 1995. Information technology – Generic coding of moving pictures and associated audio information: Video.

[9-3] Test-model editing committee, Test Model 5, MPEG93/457, ISO/IEC JTC1/SC29/WG11, April 1993.

[9-4] J. G. Kim, H. Lee, J. H. Jeong and S. Park, "MPEG-2 Compliant Transport Stream Generation; A Computer Simulation Approach." Proceedings, JCSPAT'97, San Diego, pp.152-156, Oct.1997.

[9-5] D. K. Pibush, "Timing and Synchronization Using MPEG-2 Transport Streams," SMPTE Journal, pp.395-400, July 1996.

[9-6] Jae-Gon Kim and J. Kim, "Design of a jitter-free Transport Stream Multiplexer for DTV/HDTV simulcast", Proceedings, JCSPAT'96, Boston, pp. 122-126, Oct.1996.

[9-7] J. G. Kim, H. Lee, J. Kim, and J. H. Jeong, "Design and implementation of an MPEG-2 transport stream multiplexer for HDTV satellite broadcasting", IEEE Transactions on Consumer Electronics, vol. 44, no. 3, August 1998.

[9-8] Xuemin Chen and Fan Ling, "Implementation architectures of a multi-channel MPEG-2 video transcoder using multiple programmable processors", US Patent No. 6275536B1, Aug. 14, 2001.

[9-9] Xuemin Chen, Fan Lin, and Ajay Luthra, "Video rate-buffer management scheme for MPEG transcoder", WO0046997, 2000.
[9-10] D. H. Gardner, J. E. Kaye, P. Haskell, "Remultiplexing variable rate-bitstreams using a delay buffer and rate estimation", US Patent No. 6327275, 2001.

A Basics on Digital Video Transmission Systems

A.1 Concept of Concatenated Coding System

One of the goals of channel coding research is to find a class of codes and associated decoders such that the probability of error could be made to decrease exponentially at all rates less than channel capacity while the decoding complexity increased only algebraically. Thus the discovery of such codes would make it possible to achieve an exponential tradeoff of performance vs. complexity.

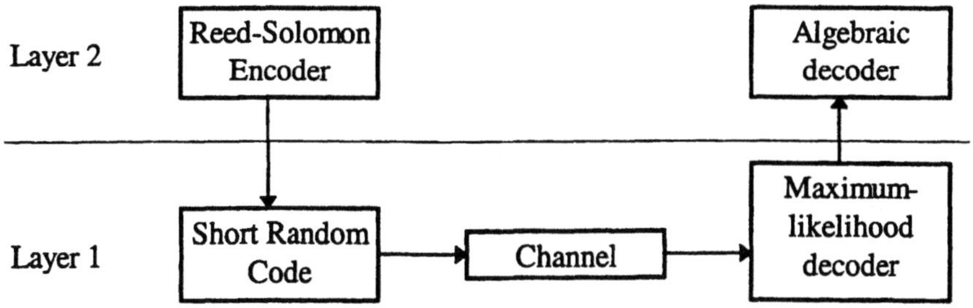

Figure A.1 A Concatenated Coding

One solution to this quest is called concatenated coding. Concatenated coding has the multilevel coding structure [A-1], illustrated in Figure A.1. In the lowest physical layer of a data network, a relatively short random "inner code" can be used with maximum-likelihood decoding to achieve a modest error probability, say, a bit-error rate of $P_b \approx 10^{-2}$, at a code-rate that is near channel capacity. Then in a second layer, a long high-rate algebraic non-binary Reed-Solomon (RS) "outer code" can be used along with a powerful algebraic error-correction algorithm to drive down the error probability to a level as low as desired with only a small code-rate loss.

RS codes have a number of characteristics that make them quite popular. First of all they have a very efficient bounded-distance decoding algorithms such as the Berlekamp-Massey algorithm or the Euclidean algorithm [A-4]. Being non-binary, RS codes also provide a significant burst-error-correcting capability. Perhaps the only disadvantage in using RS codes lies in their lack of an efficient maximum-likelihood soft-decision decoding algorithm. The difficulty in finding such an algorithm is in part due to the mismatch between the algebraic structure of a finite field and the real-number values at the output of receiver demodulator.

In order to support reliable transmission over a Gaussian channel with a binary input, it is well-known that the required minimum $\frac{E_b}{N_0}$ is - 1.6 dB for soft-decision decoders, which increases to 0.4 dB for hard-decision decoders. Here $\frac{E_b}{N_0}$ is the ratio of the received energy per information bit to the one-sided noise power spectral density. For binary block codes the above result assumes that the code-rate approaches zero asymptotically with code length. For a rate-1/2 code the minimum $\frac{E_b}{N_0}$ necessary for reliable transmission is 0.2 dB for soft-decision decoders and 1.8 dB for hard-decision decoders. These basic results suggest the significant loss of performance when soft-decision decoding is not available for a given code.

The situation is quite different for convolutional codes that use Viterbi decoding. Soft decisions are incorporated easily into the Viterbi decoding algorithm in a very natural way, providing an increase in coding gain of over 2.0 dB with respect to the comparable hard-decision decoder over an additive white Gaussian noise channel. Unfortunately convolutional codes present their own set of problems. For example, they cannot be implemented easily at high coding rates. They also have an unfortunate tendency to generate burst errors at the decoder output as the noise level at the input is increased.

A "best-of-both-worlds" situation can be obtained by combining RS codes with convolutional codes in a concatenated system. The convolutional code (with soft-decision Viterbi decoding) is used to "clean up" the channel for the Reed-Solomon code, which in turn corrects the burst errors emerging from the Viterbi decoder. Therefore, by the proper choice of codes the probability of error can be made to decrease exponentially with overall code length at all rates less than capacity. Meanwhile, the decoding complexity is dominated by the complexity of the algebraic RS decoder, which increases only algebraically with the code length.

Generally, the "outer" code is more specialized in preventing errors generated by the "inner" code when it makes a mistake. The "inner" code can also be a binary block code other than a binary convolutional code. For a band-limited channel, trellis codes often are selected as "inner" codes. To further improve error-correction performance, interleavers are usually applied between the "inner" and "outer" codes to provide resistance to burst errors.

In the next section, the state-of-the-art in concatenated coding systems is demonstrated in a video application.

A.2 Concatenated Coding Systems with Trellis Codes and RS Codes

In many digital video applications, the data format input to the modulation and channel coding is an MPEG-2 transport, as defined in reference [A-2]. Here the MPEG-2 transport is a 188-byte data-packet assembled from compressed video and audio bit-streams.

As an example, Figure A.2 shows a simplified block diagram of digital video transmission over cable networks [A-8].

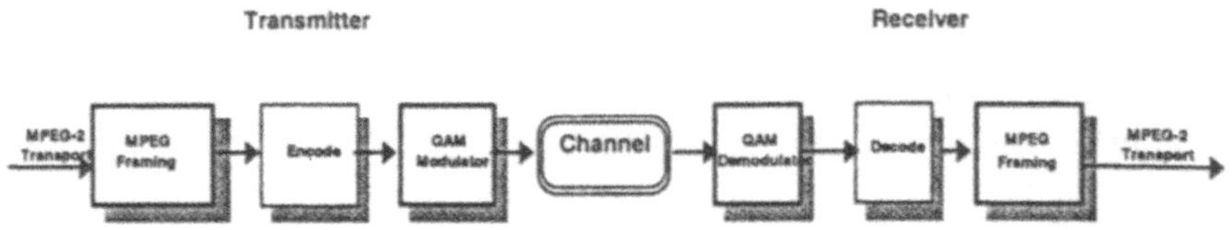

Figure A.2. Digital video transmission over cable networks

Channel coding and transmission are specific to a particular medium or communication channel. The expected channel-error statistics and distortion characteristics are critical in determining the appropriate error correction and demodulation. The cable channel, including fiber trucking, is primarily regarded as a bandwidth-limited linear channel with a balanced combination of white noise, interference, and multi-path distortion. The design of the modulation, interleaving, and channel coding is based on the testing and characterization of transmission systems. The (channel) encoding is based on a concatenated coding approach that produces high coding gains at a moderate complexity and overhead. Concatenated coding offers improved performance over a block code with a similar overall complexity. The concatenate coding system can be optimized for an almost error-free operation, for example, at a threshold output error-event rate of one error-event per 15 minutes [A-3]. The Quadrature Amplitude Modulation (QAM) technique, together with concatenated coding, is well suited to this application and channel.

In this section only channel coding blocks are discussed. The channel coding is composed of four processing layers. As illustrated in Figure A.3, the channel coding uses various types of error correcting algorithms and deinterleaving techniques to transport data reliably over the cable channel.

- **RS Coding** – Provides block encoding and decoding to correct up to three symbols within an RS block.
- **Interleaving** – Evenly disperses the symbols. This is applied for protecting against a burst of symbol errors from being sent to the RS decoder.
- **Randomization** – Randomizes the data on the channel to allow effective QAM demodulator synchronization.
- **Convolutional Coding** – Provides convolutional encoding and soft decision trellis decoding of random channel errors.

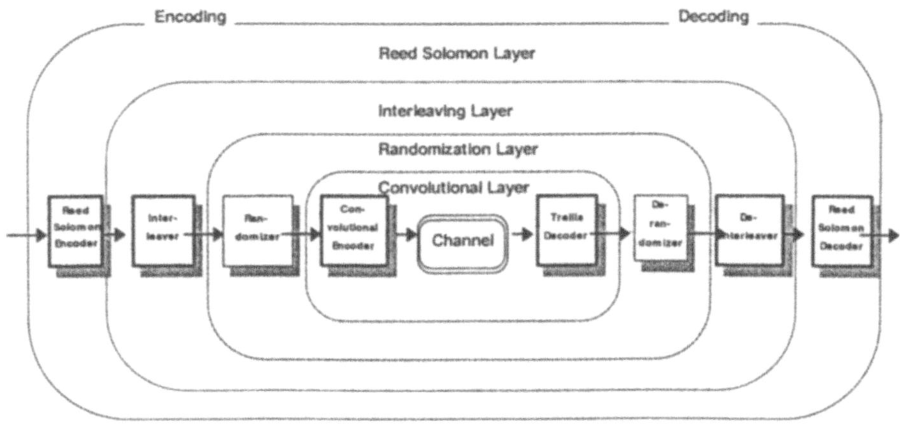

Figure A.3. Layers of Channel Coding

RS Coding: The data stream (MPEG-2 transport, etc.) is Reed-Solomon encoded using a (128,122) code over GF(2^7). This code has the capability of correcting up to t=3 errors per RS block.

The Reed-Solomon encoder is implemented as follows: A systematic encoder is utilized to implement a t=3, (128,122) extended Reed Solomon code over GF(128). The primitive polynomial used to form the field over GF(128) is: $p(x) = x^7 + x^3 + 1$.

The generator polynomial used by the encoder is:

$$g(x) = (x+\alpha)(x+\alpha^2)(x+\alpha^3)(x+\alpha^4)(x+\alpha^5)$$
$$= x^5 + \alpha^{52}x^4 + \alpha^{116}x^3 + \alpha^{119}x^2 + \alpha^{61}x + \alpha^{15}.$$

The message polynomial input to the encoder consists of 122, 7-bit symbols, and is described as follows:

$$m(x) = m_{121}x^{121} + m_{120}x^{120} + \dots m_1x + m_0.$$

This message polynomial is first multiplied by x^5, then divided by the generator polynomial g(x) to form a remainder, described by the following:

$$r(x) = r_4x^4 + r_3x^3 + r_2x^2 + r_1x + r_0.$$

This remainder constitutes five parity symbols which are then added to the message polynomial to form a 127-symbol code word that is an even multiple of the generator polynomial..

The generated code word is now described by the following polynomial:

$$c(x) = m_{121}x^{126} + m_{120}x^{125} + m_{119}x^{124} + \ldots$$
$$+ r_4x^4 + r_3x^3 + r_2x^2 + r_1x + r_0$$

By construction a valid code word has roots at the first through fifth powers of the primitive field element *a*.

An extended parity symbol c_p is generated by evaluating the code word at the sixth power of alpha as

$$c_p = c(\alpha^6)$$

This extended symbol is used to form the last symbol of a transmitted RS codeword. The extended code word then appears as follows:

$$\bar{c}(x) = m_{121}x^{127} + m_{120}x^{126} + \ldots + r_1x^2 + r_0x + c_p$$

The structure of the RS codeword that illustrates the order in which the symbols are transmitted from the output of the RS encoder is shown as follows:

$$m_{121} \; m_{120} \; m_{119} \; \cdots \; r_1 \; r_0 \; c_p$$

Note that the order that symbols are sent is from left to right.

Interleaving: Interleaving is included in the modem between the RS block coding and the randomizer to enable the correction of burst-noise-induced errors. A convolutional interleaver with depth I=128 field symbols is employed.

Convolutional interleaving is illustrated in Figure A.4. The interleaving commutator position is incremented at the RS symbol frequency, with a single symbol output from each position. In the convolutional interleaver the R-S code symbols are sequentially shifted into a bank of 128 registers. Each successive register has M-symbols more storage than the preceding register. The first interleaver path has zero delay, the second has a M symbol period of delay, the third 2*M-symbol period of delay, and so on, up to the 128th path which has 127*M-symbol period of delay. This is reversed for the deinterleaver in the Cable Decoder in such a manner that the net delay of each RS symbol is the same through the interleaver and deinterleaver. Burst noise in the channel causes a series of incorrect symbols. These are spread over many RS codewords by the deinterleaver in such a manner that the resultant in a symbol errors per codeword are within the range of the RS decoder-correction capability.

Randomization: The randomizer is the third layer of processing in the FEC block diagram. The randomizer provides for even distribution of the symbols in the constellation, which enables the demodulator to maintain proper lock. The randomizer adds a pseudo-random noise (PN) sequence to the transmitted signal to assure a random transmitted sequence.

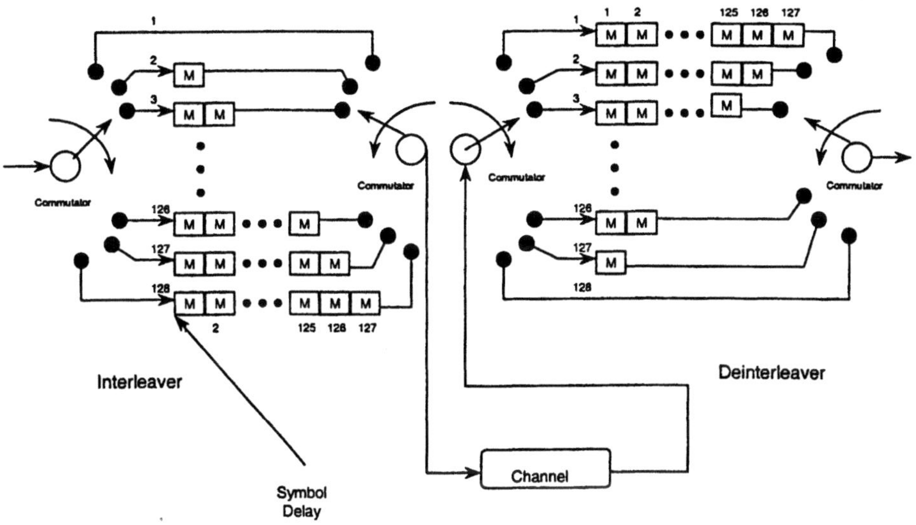

Figure A.4. Interleaving Functional Block Diagram

Trellis Coded Modulation: As part of the concatenated coding scheme, trellis coding is employed as the inner code. It allows for the utilization of redundancy to improve the Signal to Noise Ratio (SNR) by increasing the size of the symbol constellation without increasing the symbol rate. As such, it is more properly termed "trellis-coded modulation". Some basics on modulation techniques are provided in the next section.

The trellis-coded modulator includes a binary convolutional encoder to provides the appropriate SNR gain. Figure A.5 shows a 16-state non-systematic rate-1/2 convolutional encoder. The outputs of the encoder are fed into the puncturing matrix that essentially converts the rate-1/2 encoder to rate-4/5 encoder.

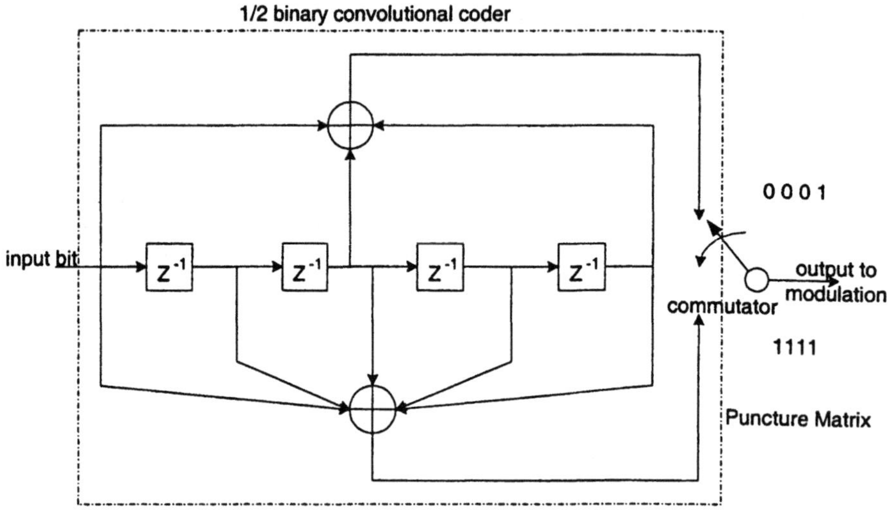

Figure A.5. Punctured binary-convolutional coder

A.3. Some Basics on Transmitter and Receiver

The process of sending (information) messages from a transmitter to a receiver is essentially a random experiment. The transmitter selects one message and sends it to the receiver. The receiver has no knowledge about which message is chosen by the transmitter, for if it did, there would be no need for the transmission. The transmitted message is chosen from a set of messages known to the transmitter and the receiver. If there were no noise, the receiver could identify the message by searching through the entire set of messages. However, the transmission medium, called the channel, usually adds noise to the message. This noise is characterized as a random process. In fact thermal noise is generated by the random motion of molecules or particles in the receiver's signal-sensing devices. Most noise has the property that it adds linearly to the received signal.

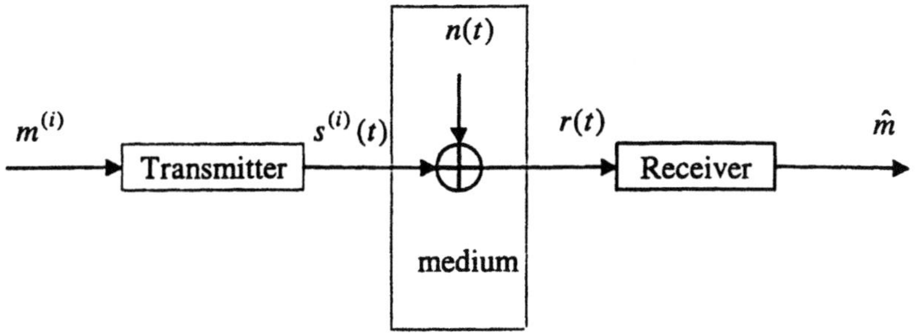

Figure A.6 A simplified block diagram of a communication process over a channel with additive noise.

Figure A.6 shows a simplified system block diagram of a transmitter/receiver communication system. The transmitter performs the random experiment of selecting one of the M messages in the message set $\{m^{(i)}\}$, say $m^{(i)}$, and then sends it corresponding waveform $s^{(i)}(t)$, chosen from a set of signals $\{s^{(i)}(t)\}$. A large body of literature is available on how to model channel impairments of the transmitted signal. This book concentrates primarily on the almost ubiquitous case of additive white-Gaussian noise (AWGN).

A.3.1 Vector Communication Channels

One commonly used method to generate signals at the transmitter is to synthesize them as a linear combination of *N* basis waveforms $\phi_j(t)$. That is, the transmitter selects

$$s^{(i)}(t) = \sum_{j=1}^{N} s_j^{(i)} \phi_j(t) \tag{A.1}$$

as the transmitted signal for the *i*-th message. Often the basis waveforms are chosen to be orthonormal; that is, they fulfill the condition,

$$\int_{-\infty}^{\infty} \phi_u(t)\phi_v(t)dt = \delta_{jl} = \begin{cases} 1, & u = v, \\ 0, & otherwise. \end{cases} \tag{A.2}$$

This leads to a vector interpretation of the transmitted signals, since, once the basis wav6forms are specified, $s^{(i)}(t)$ is completely determined by the N-dimensional vector,

$$\bar{s}^{(i)} = \left(s_1^{(i)}, s_2^{(i)}, \ldots, s_N^{(i)}\right). \tag{A.3}$$

These signals can be visualized geometrically as the signal vectors $\bar{s}^{(i)}$ in the Euclidean N-space, spanned by the usual orthonormal basis vectors, where each basis vector is associated with a basis function. This geometric representation of a signal is called a signal constellation. The idea is illustrated for $N = 2$ in Figure A.7 for the signals $s^{(1)}(t) = \sin(2\pi f_0 t)u(t)$, $s^{(2)}(t) = -\cos(2\pi f_0 t)u(t)$, $s^{(3)}(t) = -\sin(2\pi f_0 t)u(t)$,and $s^{(2)}(t) = -\cos(2\pi f_0 t)u(t)$ where

$$u(t) = \begin{cases} \sqrt{\dfrac{2E_s}{T}}, & 0 \le t \le T, \\ 0, & otherwise, \end{cases} \tag{A.4}$$

and $f_0 = \dfrac{k}{T}$ is an integer multiple of $\dfrac{1}{T}$. The first basis function is $\phi_1(t) = \sqrt{\dfrac{2}{T}} \sin(2\pi f_0 t)$, and the other basis function is $\phi_2(t) = \sqrt{\dfrac{2}{T}} \cos(2\pi f_0 t)$. The signal; constellation in Figure A.7 is called quadrature phase-shift keying (QPSK).

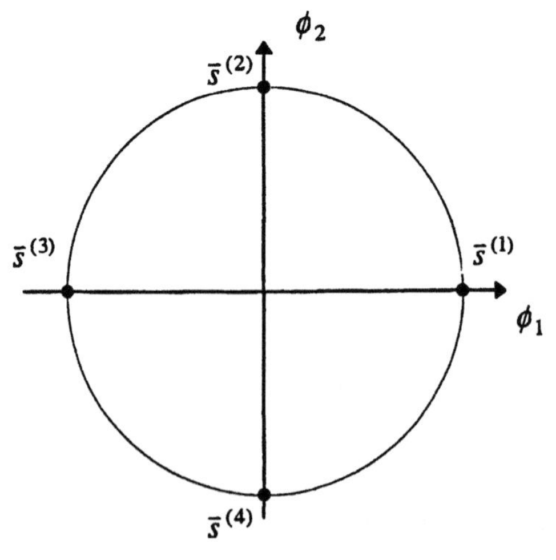

Figure A.7 Illustration of QPSK

There is a one-to-one mapping at the signal vector $\bar{s}^{(i)}$ onto the transmitted message $\bar{m}^{(i)}$. The problem of decoding a received waveform is therefore equivalent to the ability to recover the signal vector $\bar{s}^{(i)}$. This can be accomplished by passing the received signal waveform through a bank of correlators where each correlator correlates $s^{(i)}(t)$ with one of the basis functions to perform the operation,

$$\int_{-\infty}^{\infty} s^{(i)}(t)\phi_j(t)dt = \sum_{l=1}^{N} s_l^{(i)} \int_{-\infty}^{\infty} \phi_l(t)\phi_j(t)dt = s_j^{(i)} . \tag{A.5}$$

That is, the j-th correlator recovers the j-th component $s_j^{(i)}$ of the signal vector $\bar{s}^{(i)}$.

Next, define the squared Euclidean distance between two signals $\bar{s}^{(i)}$ and $\bar{s}^{(j)}$, given by $d_E^2(\bar{s}^{(i)},\bar{s}^{(j)}) = \left|\bar{s}^{(i)} - \bar{s}^{(j)}\right|^2$, which is a measure of, what is called, the noise resistance of these two signals. Furthermore, the expression,

$$d_E^2(\bar{s}^{(i)},\bar{s}^{(j)}) = \sum_{l=0}^{N}\left(s_l^{(i)} - s_l^{(j)}\right)^2 = \int_{-\infty}^{\infty}\left(s^{(i)}(t) - s^{(j)}(t)\right)^2 dt \tag{A.6}$$

is, in fact, the energy of the difference signal $s^{(i)}(t) - s^{(j)}(t)$.

It can be shown that the correlation-receiver is optimal, in the sense that no relevant information is discarded and that the minimum error probability is attained, even when the received signal contains additive white Gaussian noise. In this latter case the received signal $r(t) = s^{(i)}(t) + n(t)$ produces the received vector $\bar{r} = \bar{s}^{(i)} + \bar{n}$ at the output of the bank of correlators. The statistics of the noise vector $\bar{n}$ are easily evaluated, using the orthogonality of the basis waveforms and the noise correlation function $E[n(t)n(t+\tau)] = \frac{N_0}{2}\delta(t)$ for white-Gaussian noise, where $\delta(t)$ is Dirac's delta function and N_0 is the one-sided noise power spectral density. Thus the correlation of any two noise components n_u and n_v is given by

$$E[n_u n_v] = \int_{-\infty}^{\infty}\int_{-\infty}^{\infty} E[n(\alpha)n(\beta)]\phi_u(\alpha)\phi_v(\beta)d\alpha d\beta = \begin{cases} \frac{N_0}{2}, & u = v, \\ 0, & otherwise. \end{cases} \tag{A.7}$$

It is seen, that by the use of the orthonormal basis waveforms, the components of the random noise vector n are all uncorrelated. Since $n\ (t)$ is a Gaussian random process, the sample or component values n_u, n_v are necessarily also Gaussian. From the foregoing one concludes that the components of n are independent, Gaussian random variables with a common variance $N_0/2$ and zero mean value.

The advantages of the above vector point of view are manifold. First, one doesn't need to be concerned with the actual choices of the signal waveforms when discussing receiver algorithms. Secondly, the difficult problem of waveform communication, involving stochastic processes and continuous signal functions, has been transformed into the much more manageable vector communications system which involves only signal and random vectors.

A.3.2 Optimal Receivers

If the bank of correlators produces a received vector $\bar{r} = \bar{s}^{(i)} + \bar{n}$, then an optimal detector chooses that message hypothesis $\hat{m} = \bar{m}^{(j)}$ which maximizes the conditional probability $P\left[\bar{m}^{(j)} | \bar{r}\right]$. This is known as a maximum aposteriori (MAP) receiver.

Evidently, a use of Bayes' rule yields

$$P\left[\bar{m}^{(j)} | r\right] = \frac{P[\bar{m}]p\left(\bar{r} | \bar{m}^{(j)}\right)}{p(\bar{r})}. \tag{A.8}$$

Thus, if all of the signals are used equally often, the maximization of $P\left[\bar{m}^{(j)} | \bar{r}\right]$ is equivalent to the maximization of $p(\bar{r} | \bar{m}^{(j)})$. This is the maximum-likelihood (ML) receiver. It minimizes the signal-error probability only for equally-likely signals.

Since $\bar{r} = \bar{s}^{(i)} + \bar{n}$ and $\bar{n}$ is an additive Gaussian random vector which is independent of the signal $\bar{s}^{(i)}$, the optimal receiver is derived by the use of the conditional probability density $p\left(\bar{r} | \bar{m}^{(j)}\right) = p_n\left(\bar{r} - \bar{s}^{(j)}\right)$. Specifically, this is the N-dimensional Gaussian density function given by

$$p_n\left(\bar{r}-\bar{s}^{(j)}\right)=\frac{1}{\left(\pi N_0\right)^{N/2}}\exp\left(-\frac{\left|\bar{r}-\bar{s}^{(j)}\right|^2}{N_0}\right). \tag{A.9}$$

The maximization of $p(\bar{r}|\bar{m}^{(j)})$ is seen to be equivalent to the minimization of the squared-Euclidean distance,

$$d_E^2\left(\bar{r},\bar{s}^{(j)}\right)=\left|\bar{r}-\bar{s}^{(j)}\right|^2, \tag{A.10}$$

between the received vector and the hypothesized signal vector.

The decision rule in (A.10) implies that the decision regions $D^{(j)}$ for each signal point consists of all the points in Euclidean N-dimensional space that are closer to $\bar{s}^{(j)}$ than any other signal point. Such decision regions for QPSK are illustrated in Figure A.8.

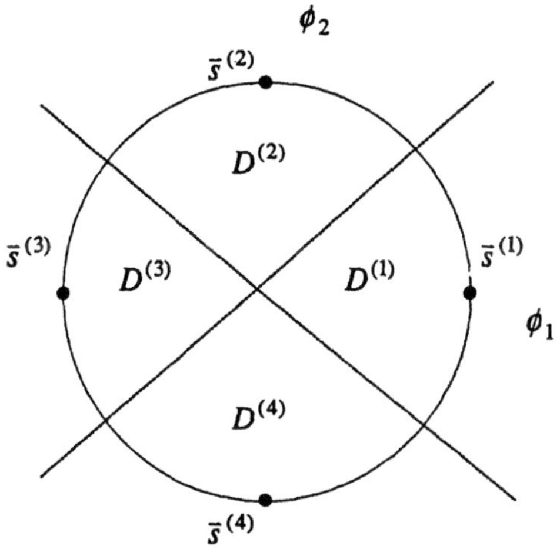

Figure A.8 Decision regions for ML detection of QPSK signals.

Hence the probability of error, given a particular transmitted signal $\bar{s}^{(i)}$, can be interpreted as the probability that the additive noise $\bar{n}$ carries the signal $\bar{s}^{(i)}$ outside its decision region $D^{(i)}$. This probability is calculated by

$$P\left(error|\bar{s}^{(i)}\right)=\int_{\bar{r}\in D^{(i)}} p_n(\bar{r}-\bar{s}^{(i)})d\bar{r}\,. \tag{A.11}$$

Equation (A.11) is, in general, quite difficult to calculate in closed form, and simple expressions exist only for certain special cases. The most important such special case is the two-signal error probability. This is the probability that signal $\bar{s}^{(i)}$ is decoded as signal $\bar{s}^{(j)}$ on the assumption that there are only these two signals. To calculate the two-signal error probability all signals are disregarded except $\bar{s}^{(i)}$, $\bar{s}^{(j)}$, e.g. $\bar{s}^{(i)} = s^{(1)}$ and $\bar{s}^{(j)} = s^{(3)}$ in Figure A.8. The new decision regions are $D^{(i)} = D^{(1)} \cup D^{(4)}$ and $D^{(j)} = D^{(2)} \cup D^{(3)}$. The decision region $D^{(i)}$ is expanded to the half-plane, and the probability of deciding on message $\bar{m}^{(j)}$ when message $\bar{m}^{(i)}$ was actually transmitted is

$$P_{\bar{s}^{(i)} \to \bar{s}^{(j)}} = Q\left(\sqrt{\frac{d_E^2\left(\bar{s}^{(i)}, \bar{s}^{(j)}\right)}{2N_0}}\right) \tag{A.12}$$

where $d_E^2\left(\bar{s}^{(i)}, \bar{s}^{(j)}\right) = \left|\bar{s}^{(i)} - \bar{s}^{(j)}\right|^2$ is the energy of the difference signal, and

$$Q(u) = \frac{1}{\sqrt{2\pi}} \int_u^\infty \exp\left(-\frac{v^2}{2}\right) dv \tag{A.13}$$

is a nonelementary integral, called the (Gaussian) Q-function. The probability in (A.12) is known as the *pairwise error probability*.

The correlation operation in (A.5), used to recover the signal-vector components, can be implemented as a filtering operation. The signal $s^{(i)}(t)$ is passed through a filter with time impulse response $\phi_u(-t)$ to obtain

$$x(t) = s^{(i)}(t) * \phi_u(-t) = \int_{-\infty}^{\infty} s^{(i)}(y)\phi_u(y-t)\,dy\,. \tag{A.14}$$

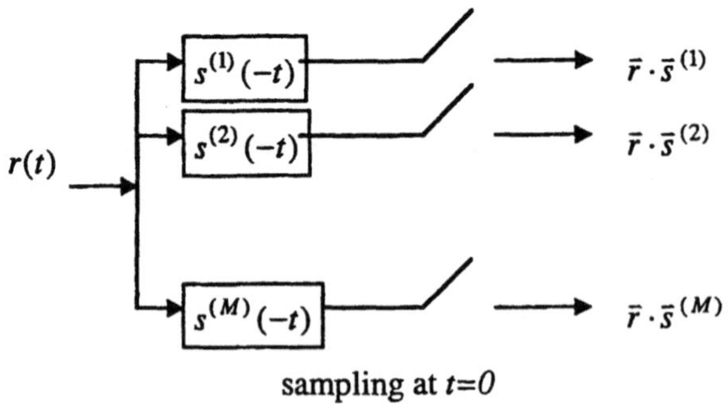

Figure A.9. Matched filter receiver, where M is the number of messages.

If the output of the filter $\phi_u(-t)$ is sampled at time t = 0, equations (A.14) and (A.5) are identical; i.e.,

$$x(t=0) = \int_{-\infty}^{\infty} s^{(i)}(y)\phi_u(y)dy \tag{A.15}$$

Of course, some appropriate delay actually needs to be built into the system in order to guarantee that $\phi_u(-t)$ is a causal filter. Such delays are not considered further in here.

The maximum-likelihood receiver minimizes $d_E^2\left(\bar{r},\bar{s}^{(i)}\right)=\left|\bar{r}-\bar{s}^{(i)}\right|^2$ or, equivalently, it maximizes

$$2\cdot\bar{r}\cdot\bar{s}^{(i)}-\left|\bar{s}^{(i)}\right|^2, \tag{A.16}$$

where the term $\left|\bar{r}\right|^2$, which is common to all the hypotheses, is neglected. The correlation $\bar{r}\cdot\bar{s}^{(i)}$ is the central part of (A.16) and can be implemented as a basis-function matched-filter receiver, where the summation is performed after the correlation. That is,

$$\bar{r}\cdot\bar{s}^{(i)}=\sum_{j=1}^{N} r_j s_j^{(i)}=\sum_{j=1}^{N} s_j^{(i)}\int_{-\infty}^{\infty} r(t)\phi_j(t)dt \tag{A.17}$$

Such a receiver is illustrated in Figure A.9. Usually the number of basis functions is much smaller than the number of signals so that the basis-function matched-filter implementation is the preferred realization.

A.3.3 Message Sequences

In practice, information signals $s^{(i)}(t)$ will most often consist of a sequence of identical, time-displaced waveforms, called pulses, described by

$$s^{(i)}(t)=\sum_{j=-u}^{u} a_j p(t-jT), \tag{A.18}$$

where *p(t)* is some pulse waveform, the a_j are the discrete symbol values from some finite signal alphabet (e.g., the binary signaling: $a_j\in\{-1,1\}$), and $2u+1$ is the length of the sequence of symbols. The parameter *T* is the timing delay between successive pulses, also called the *symbol period.* The output of the filter matched to the signal $s^{(i)}(t)$ is given by

$$y(t) = \sum_{j=-u}^{u} a_j \int_{-\infty}^{\infty} r(v)p(v - jT - t)dv, \tag{A.19}$$

and the sampled value of $y(t)$ at $t=0$ is given by

$$y(t = 0) = \sum_{j=-u}^{u} a_j \int_{-\infty}^{\infty} r(v)p(v - jT)dv = \sum_{j=-u}^{u} a_j y_j \tag{A.20}$$

where $y_j = \int_{-\infty}^{\infty} r(v)p(v - jT)dv$ is the output of the filter which is matched to the pulse $p(t)$ sampled at time t = jT. Thus the matched filter can be implemented by the pulse-matched filter, whose output $y(t)$ is sampled at multiples of the symbol time T.

In many practical applications one needs to shift the center frequency of a narrowband signal to some higher frequency band for purposes of transmission. The reason for this may lie in the transmission properties of the physical channel, which allows the passage of signals only in some unusually high-frequency bands. This occurs, for example, in radio transmission. The process of shifting a signal in frequency is called modulation by a carrier frequency. Modulation is also important for wire-bound transmissions, since it makes possible the coexistence of several signals on the same physical medium, all residing in different frequency bands; this is known as frequency division multiplexing (FDM). Probably the most popular modulation method for digital signals is quadrature double-sideband suppressed- carrier (DSB-SC) modulation.

DSB-SC modulation is a simple linear shift in frequency of a signal $x(t)$ with low-frequency content, called baseband, into a higher-frequency band by multiplying $x(t)$ by a cosine or sine waveform with carrier frequency f_0, as shown in Figure A.10, to obtain the signal,

$$s_0(t) = x(t)\cos(2\pi f_0 t), \tag{A.21}$$

on carrier to, where the factor $\sqrt{2}$ is used to make the powers of $s_0(t)$ and $x(t)$ equal.

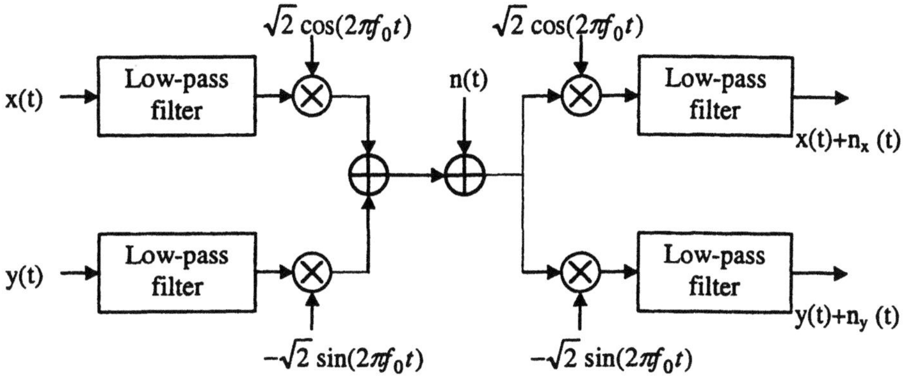

Figure A.10 Quadrature DSB-SC modulation and demodulation stages.

If the baseband signal x(t) occupies frequencies which range from 0 to *W* Hz, then $s_0(t)$ occupies frequencies from $f_0 - W$ to $f_0 + W$ Hz, an expansion of the bandwidth by a factor of 2. But we quickly note that another signal, $-y(t)\sqrt{2}\sin(2\pi f_0 t)$, can be put into the same frequency band and that both baseband signals *x(t)* and *y(t)* can be recovered by the demodulation operation shown in Figure A.10. This is the product demodulator, where the low-pass filters *W(f)* serve to reject unwanted out-of-band noise and signals. It can be shown that this arrangement is optimal. That is, no information or optimality is lost by using the product demodulator for DSB-SC-modulated signals.

If the synchronization between the modulator and demodulator is perfect, the signals *x(t)*, the *in-phase* signal, and *y(t)*, the *quadrature* signal, are recovered independently without either affecting the other. The DSB-SC-modulated bandpass channel is then, in essence, a dual channel for two independent signals, each of which may carry an independent data stream.

In view of our earlier approach that used basis functions one may want to view each-pair of identical input pulses to the two channels as a two-dimensional signal. Since these two dimensions are intimately linked through the carrier modulation, and since bandpass signals are so ubiquitous in digital communications, a complex notation for bandpass signals has been widely adopted. In this notation, the in-phase signal *x(t)* is real, and the quadrature signal *jy(t)* is an imaginary signal, expressed by

$$
\begin{aligned}
s_0(t) &= x(t)\sqrt{2}\cos(2\pi f_0 t) - y(t)\sqrt{2}\sin(2\pi f_0 t) \\
&= \mathrm{Re}\left[(x(t) + jy(t))\sqrt{2}e^{2\pi j f_0 t}\right],
\end{aligned}
\tag{A.22}
$$

where $s(t) = x(t) + jy(t)$ is called the complex envelope of $s_0(t)$.

Bibliography

[A.1] G. D. Forney, Jr., Concatenated Codes, Cambridge, MA : MIT Press, 1966.

[A-2] ITU-T Recommendation H.222.0 (1995) | ISO/IEC 13818-1 : 1996, Information technology--generic coding of moving pictures and associated audio information systems.

[A-3] Prodan, R. et al., "Analysis of Cable System Digital Transmission Characteristics," NCTA Technical Papers, 1994.

[A-4] Irving S. Reed, and Xuemin Chen, Error-Control Coding for Data Networks, 2nd Print, Kluwer Academic Publishers, Boston, 2001.

[A-5] IEEE Project 802.14/a, "Cable-TV access method and physical layer specification", 1997.

[A-6] S. Lin and D. Costello, Error Control Coding: *Fundamentals and Applications,* Englewood Cliffs, NJ: Prentice Hail, Inc., 1983.

[A-7] J. Hagenauer, E. Offer, and L. Papke, "Matching Viterbi decoders and Reed-Solomon decoders in a concatenated system," in Reed Solomon Codes and Their Applications, New York : IEEE Press, 1994.

[A-8] ITU-T Telecommunication Standardization Sector of ITU, "Digital multi-programmer systems for television sound and data services for cable distribution"—Television and sound transmission, ITU-T Recommendation J.83, Oct. 1995.

[A-9] O. M. Collins and M. Hizlan, "Determinate state convolutional codes", IEEE Trans. Communications, vol. 41, pp.1785-1794, Dec. 1993.

[A-10] J. Hagenauer and P. Hoeher, "A Viterbi algorithm with soft-decision outputs and its applications," in Proc. 1989 IEEE Global Communication Conference, pp.47.1.1-47.1.7, Dallas, TX, Nov. 1989.

[A-11] S. Proalds, Digital Communications, third edition. New York: McGraw-Hill, Inc., 1995.

[A-12] S. Wicker, Error Control Systems for Digital Communication and Storage. Englewood Cliffs, NJ: Prentice Hall, Inc., 1995.

Index

A

B

C

D

E

Xuemin Chen has more than 15 years experience on broadband communication system architectures, digital video and television transmission systems, and media-processor/DSP/ASIC architectures. He is a Senior Member of IEEE and has a Ph. D. degree in Electrical Engineering from University of Southern California (USC). He co-authored (with Prof. Irving S. Reed) a graduate-level textbook, entitled "Error-Control Coding for Data Networks" (Kluwer Academic Publishers, 1st print 1999, 2nd print 2001).

Dr. Chen is the inventor of more than 40 granted or published patents worldwide in digital image/video processing and communication. He has also published over 60 research articles and contributed many book chapters in data compression and channel coding.

Dr. Chen has made many significant contributions in the architecture design and system implementation of digital video communication systems/chips. He also actively involved in developing ISO/IEC MPEG-2 and MPEG-4 standards.

GPSR Compliance

The European Union's (EU) General Product Safety Regulation (GPSR) is a set of rules that requires consumer products to be safe and our obligations to ensure this.

If you have any concerns about our products, you can contact us on ProductSafety@springernature.com

In case Publisher is established outside the EU, the EU authorized representative is:

Springer Nature Customer Service Center GmbH
Europaplatz 3
69115 Heidelberg, Germany

Batch number: 10620086

Printed by Printforce, the Netherlands